First published in 2020 by Barrallier Books Pty Ltd,
trading as Echo Books

Registered Office: 35—37 Gordon Avenue, West Geelong, Victoria 3220, Australia.

www.echobooks.com.au

Copyright ©Marcus Fielding

Creator: Fielding, Marcus.

Title: Dealing with a Deadly Legacy: Aussie Soldiers Clearing Land Mines in Afghanistan

ISBN: 978-0-6485540-9-7 (softcover)

Book layout and design by Peter Gamble, Canberra.
Set in Garamond Premier Pro Display, 12/17 MinervaSculptura, MinervaSmallcaps.

www.echobooks.com.au

DEALING
with a
DEADLY
LEGACY

Aussie Soldiers Clearing Land Mines
in Afghanistan

Marcus Fielding

ECHO BOOKS

Darrell Crichton sporting a beard and a koala at Peshawar in 1992.

This book is dedicated to those Australian Army
United Nations Mine Clearance Training Team members
who have since died:

Ross John Chamberlain
Brian Clegg
Darrell William Crichton
Kevin Desmond Darcy
Brian Edward Gardner
Danny John Hawkins
James Charlton Horton
George Patrick O'Callaghan
Phillip James Palazzi
Donald Alan Quick
George James Turner

CONTENTS

Foreword

By Professor William Maley

In June 1993, I found myself in the Pakistani city of Peshawar, which I had been visiting regularly for some years to conduct research on Afghan refugees, and on the evolving situation in Afghanistan beyond the 'Durand Line' which had been drawn in 1893 to separate Afghanistan from British India. Out of the blue, I received a call from an Australian army officer, deployed as part of the United Nations Mine Clearance Training Team (UNMCTT), who wanted urgently to meet. I arranged to catch up with him and one of his colleagues on 5 June, and they briefed me on two matters. One was the work that the UNMCTT had been conducting in order to try to address the frightful problem of landmine contamination that Afghanistan faced as a consequence of the years of conflict between the Soviet invasion of Afghanistan in December 1979, and the withdrawal of Soviet forces in February 1989, followed by the collapse in April 1992 of the Communist regime which those forces had been supporting. The other, which clearly had left them shaken and frustrated, was a decision taken in Canberra to withdraw the Australian contingent that had been serving with the UNMCTT. Unable themselves to speak in public about this issue, they asked me whether I could take it up in my capacity as an academic specialist on Afghanistan. I was happy to do so, and unbeknownst to me at the time, their request was to open a new vista of research and inquiry which resulted not only in my editing a short book on the landmines issue, but in years of advocacy in support of the total ban on the manufacture, stockpiling, transfer and use of antipersonnel landmines that was finally embodied in the 1997 *Ottawa Treaty*.[1] This treaty came into effect in 1999, and defined a new set of norms[2] to which many states have faithfully adhered in the decades since.

The threat to the continued work of the Australian contingent was a matter of deep concern. The United Nations had estimated in 1993 that some 10 million mines and items of unexploded ordnance (UXOs) were to be found in Afghanistan,[3] and with refugees already flowing back into the country following the collapse of the communist regime, the risk of grave injury to ordinary people from encounters with a victim-activated munition was extreme. As a result, any steps that might compromise the effectiveness of mine action in Afghanistan of the time were highly problematic. It became very clear, very quickly, that the discontinuation of the activities of the Australian contingent was not an initiative that had come from the Australian Defence Force, let alone the military personnel actually serving in the field.[4] Rather, the push to withdraw had come from a small number of civilians within

1   William Maley, 'The United Nations, NGOs and the land-mines initiative: an Australian Perspective', in Andrew F. Cooper, John English and Ramesh Thakur (eds), *Enhancing Global Governance: Towards a New Diplomacy?* (Tokyo: United Nations University Press, 2002) pp.90-105.

2   Ramesh Thakur and William Maley, 'The Ottawa Convention on Landmines: A Landmark Humanitarian Treaty in Arms Control?', *Global Governance*, vol.5, no.3, July-September 1999, pp.273-302.

3   United Nations Office for the Coordination of Humanitarian Assistance to Afghanistan, *Afghanistan: Mine Clearance Program for 1993* (Islamabad: UNOCHA, March 1993) p.2.

4   See William Maley (ed.), *Dealing with Mines: Strategies for peacekeepers, aid agencies and the international community* (Canberra: Australian Defence Studies Centre, 1994).

the Defence Department who had found something of an ally in the then Defence Minister, Senator Robert Ray, who endorsed their recommendation. In his *Official History* discussion of the issue, Professor David Horner observed that Senator Ray endorsed the proposal to withdraw knowing that 'the CGS, the Land Commander, the Department of the Prime Minister and Cabinet, the High Commissioner and the Defence Adviser in Islamabad did not agree'. Nor did Foreign Minister Senator Gareth Evans.[5] A subsequent report from the Joint Standing Committee on Foreign Affairs, Defence and Trade of the Commonwealth Parliament concluded that 'Australia's "good international citizen" image' had been 'somewhat tarnished by its withdrawal of support of UNMCTT'.[6] This unusually-sharp comment was in part a reflection of a sense on the part of the Committee that the Department of Defence had misled it about how the uniformed military actually felt about the decision. The withdrawal of the contingent sent an unfortunate signal about the scale of Australian commitment to addressing what was a global scourge, and came ironically at the very time when relations between Afghanistan and Australia were being renewed following years of inactivity on the political front.[7]

At the time that the withdrawal occurred, there was a real fear that this could lead to an unravelling of what had been an extraordinarily effective program of mine action assistance. This did not happen, but one of the reasons why it did not happen was that a number of senior and experienced staff left the ADF in order to continue to work under contract in mine-afflicted countries, including Afghanistan.[8] Ian Mansfield has written

5   David Horner, *Australia and the 'New World Order': From peacekeeping to peace enforcement: 1988-1991* (Cambridge: Cambridge University Press, 2011) p.260.

6   Joint Standing Committee on Foreign Affairs, Defence and Trade, *Australia's Participation in Peacekeeping* (Canberra: Australian Government Publishing Service, December 1994) para.4.35.

7   William Maley, *Australia-Afghanistan Relations: Reflections on a half-century* (Canberra: Australian Strategic Policy Institute, 2019).

8   Horner, *Australia and the 'New World Order'*, p.265.

a wonderful memoir that captures how rewarding it could be in a very personal sense to assist in a concrete way in addressing a problem that was blighting the lives of vulnerable people,[9] but many other Australians have similar stories to tell. This collection tells some of these stories, and in a way that demonstrates how much good can be done by those who have embraced the profession of arms. But beyond this, the book also gives a real 'smell and feel' of what soldiering can involve in an increasingly-complicated world, as well as an insight into the remarkable individuals who left Australia's shores in order to work in a quite remote and very dangerous part of the world, and in service of humanitarian objectives.

Australian personnel were to return to Afghanistan after 2001, and more than 25,000 ADF members served in operations conducted by special forces, as well as in the Provincial Reconstruction Team (PRT) to which Australia contributed in the province of Uruzgan.[10] This later deployment was to cost, conservatively, between A$12.2 billion and A$13.6 billion,[11] and much more seriously, it also cost the lives of 41 serving personnel. Afghanistan remains a troubled land. Its politics are fraught with tension,[12]

9   Ian Mansfield, *Stepping into a Minefield: A life dedicated to landmine clearance around the world* (Newport: Big Sky Publishing, 2015). See also Ian Mansfield, 'Landmines, Australians and Peacekeeping', in David Horner, Peter Londey and Jean Bou (eds), *Australian Peacekeeping: Sixty Years in the Field* (Cambridge: Cambridge University Press, 2009) pp.223-234.

10   See Chris Masters, *No Front Line: Australia's Special Forces at War in Afghanistan* (Sydney: Allen & Unwin, 2017); William Maley, 'PRT Activity in Afghanistan: The Australian Experience', in Nikola Hynek and Péter Marton (eds), *Statebuilding in Afghanistan: Multinational Contributions to Reconstruction* (New York: Routledge, 2011) pp.124-138; Karen Middleton, *An Unwinnable War: Australia in Afghanistan* (Melbourne: Melbourne University Press, 2011).

11   Peter Hall, 'The Economic Cost to Australia of the War in Afghanistan', in Jack Cunningham and William Maley (eds), *Australia and Canada in Afghanistan: Perspectives on a Mission* (Toronto: Dundurn, 2015) pp.113-131 at p.127.

12   See William Maley, *Transition in Afghanistan: Hope, Despair and the Limits of Statebuilding* (London: Routledge, 2018).

and its people continue to be threatened by Pakistan-supported extremist groups such as the Taliban,[13] as well as by radicals such as ISIS.[14] But one source of horror that haunts the daily lives of ordinary Afghans, the problem of landmines, is no longer quite so daunting. It has *not* gone away, and there are disturbing signs of US recidivism on the issue.[15] But Afghans have found ways of living more safely in a mined environment, and for this they owe a debt of gratitude to those given voice in this book.

---

13   See Antonio Giustozzi, *The Taliban at War, 2001-2018* (London: Hurst & Co., 2019).

14   See Antonio Giustozzi, *The Islamic State in Khorasan: Afghanistan, Pakistan and the New Central Asian Jihad* (London: Hurst & Co., 2018); Haroro J. Ingram, Craig Whiteside and Charlie Winter, *The ISIS Reader: Milestone Texts of the Islamic State Movement* (London: Hurst & Co., 2020).

15   Alex Horton, 'Why the land mine, a persistent killer of civilians, is coming back under Trump', *The Washington Post*, 2 February 2020.

ACRONYMS AND ABBREVIATIONS

2IC	—Second in Command
ADC	—Aide de Camp
ADF	—Australian Defence Force
ANZAC	—Australia New Zealand Army Corps
ATC	—Afghan Technical Consultants
CAPT	—Captain
CIA	—Central Intelligence Agency
CO	—Commanding Officer
CPL	—Corporal
DFAT	—Department of Foreign Affairs and Trade
DIO	—Defence Intelligence Organisation
DRA	—Democratic Republic of Afghanistan
ERW	—Explosive Remnants of War
EO	—Explosive Ordnance
EOD	—Explosive Ordnance Disposal
FATA	—Federally Administered Tribal Areas
FAO	—Food and Agriculture Organization
FBT	—Fuze Blasting Time
HQ	—Headquarters
ICRC	—International Committee for the Red Cross/Crescent
IED	—Improvised Explosive Device
IOM	—International Organisation for Migration

ISAF	—International Security Assistance Force
LTCOL	—Lieutenant Colonel
MAJ	—Major
MCPA	—Mine Clearance Planning Agency
MSF	—Medecins Sans Frontieres (Doctors Without Borders)
NATO	—North Atlantic Treaty Organisation
NGO	—Non-Government Organisation
OAM	—Order of Australia Medal
OIC	—Officer in Charge
OMA	—Organisation for Mine Awareness
PIA	—Pakistan International Airlines
PDPA	—People's Democratic Party of Afghanistan
R&R	—Rest and Recreation
RAAMC	—Royal Australian Army Medical Corps
RAE	—Royal Australian Engineers
RAInf	—Royal Australian Infantry
RPG	—Rocket Propelled Grenade
RSM	—Regimental Sergeant Major
SGT	—Sergeant
SSGT	—Staff Sergeant
SITREP	—Situation Report
SME	—School of Military Engineering
SOP	—Standard Operating Procedure
SWAAD	—South West Afghanistan Agency for Demining
UN	—United Nations
UNICEF	—United Nations Children's Fund
UNDP	—United Nations Development Program
UNGOMAP	—United Nations Good Offices Mission in Afghanistan and Pakistan
UNHCR	—United Nations High Commissioner for Refugees
UNMCTT	—United Nations Mine Clearance Training Team
UNSMA	—United Nations Special Mission to Afghanistan
UNOCA	—Office for the Coordination of United Nations Humanitarian and Economic Assistance Programmes Relating to Afghanistan

USSR	—Union of Soviet Socialist Republics
UXB	—Unexploded Bombs
UXO	—Unexploded Ordnance
WO	—Warrant Officer
WFP	—World Food Program
XO	—Executive Officer

Chapter One
Introduction

In what is probably the most extraordinary and hazardous circumstances ever faced by Australian soldiers, ninety-two combat engineers helped to clear minefields in the midst of an ongoing civil war.

Unarmed, dressed in mufti, disguised with beards and working through interpreters they helped to forge local expertise.

Adding to the risks they had only a medic on hand in the event of becoming the victim of a mine blast; and the nearest hospital was over a full day's drive away.

How none of them were killed or injured is remarkable.

These Australian Army soldiers were working as part of a United Nations humanitarian mine clearance program in Pakistan and Afghanistan between 1989 and 1993.

They blazed a path for future humanitarian land mine clearance efforts around the world.

Telling the extraordinary story of this operation and the men who participated in it is the objective of this book. The Australian Government's Official History of this operation was published in 2011 and provides a great level of detail of the government processes and decisions. But as an official document, the inclusion of more descriptive and personal stories was not appropriate. This book seeks to complement the Official History and provide a more personal (and colourful) account.

It will seek to do this by sharing memories from the participants in their own words and by providing explanation where necessary or appropriate. Coupled with photographs, maps and background 'fact boxes' it is hoped that their story can be sufficiently captured for posterity.

As each of the participants followed a common experience of being deployed and later returning to Australia the book is arranged in chapters that follow these stages; that is the chapters are titled Selection (for the deployment), Pre-deployment (training and preparation), Orientation (on arrival in Pakistan), Working (in Pakistan and in Afghanistan), Rest and Recreation, Going Home and Reflections (retrospective views on the deployment).

But first, a short summary of the UN Mine Clearance Program in Pakistan and Afghanistan between 1989 and 1993 is a good place to begin telling the story.

The UN Mine Clearance Program

The United Nations Office for the Coordination of Humanitarian Aid to Afghanistan (UNOCA) established the mine clearance training program in early 1989. The US, France and Turkey dispatched military contingents that started work at Risalpur in February 1989. The Pakistani Army Corps of Engineers provided support to the training program.

Courses were conducted for selected groups of Afghan refugees—predominantly adult males. The courses were initially three days long and focused on land mine and UXO recognition and avoidance techniques. The expectation was that refugee families would soon return to Afghanistan following the Soviet withdrawal and that these skills would enable them to enhance their safety.

Contingents from Canada, Italy and Norway joined the effort soon after, followed by New Zealand, the UK and Australia. The NZ and Australian governments referred to their contingents as being part of the UN Mine Clearance Training Team (UNMCTT)—even though the

UN used the term 'Operation Salam' to refer to the overall humanitarian program. Some national contingents were also located at a second training camp near Quetta in southern Pakistan.

The intention was for these trained personnel to voluntarily repatriate to Afghanistan and undertake mine and UXO clearance on their own initiative. Two major factors contributed to change the UN's approach to mine clearance operations in Afghanistan:

- The expectation that the Afghan refugees would return to Afghanistan *en masse*, however, proved false with the continuation of the fighting between the Soviet-backed Afghan regime with the factious Afghan Mujahideen; and

- A realisation that the socio-economic impact of land mine contamination was simply too large and complex a problem to be left to individuals and ad-hoc clearance efforts; a large scale nationally coordinated approach was needed to assist the rehabilitation strategy for Afghanistan.

Over time the training courses were extended to include techniques to safely extract people from mined areas or from mine incidents, and basic techniques to destroy land mines and UXO using explosives. The Pakistani Army and Inter-Services Intelligence would vet the students who had been nominated by one of the political parties associated with the Afghans. At the completion of the 17 days of training students were issued a canvas bag containing basic tools to perform these tasks and a basic first aid kit. Student were also required to sign an oath declaring that they would only use the knowledge and skills they had been taught for humanitarian purposes.

In November 1989 the Australian government accepted responsibility for technical advice and training of Afghan non-government organisations (NGOs) to undertake a trial of large scale and coordinated mine clearance operations under the management of the UNOCA mine clearance program. The first demining teams from Afghan Technical Consultants deployed into Afghanistan in early January 1990.

Throughout 1990 and 1991 it became clear that a large-scale centrally coordinated approach to mine clearance was feasible and security conditions in Afghanistan were sufficiently stable for UNOCA to expand on the trial of an organised demining effort. This was achieved with the formation of a number of NGOs whose actions were coordinated through regional demining offices in Peshawar, Quetta and later Kabul—the Afghan capital.

With the formal establishment of this 'Demining' (a new term) Program, Australia extended its tours for UNMCTT members to six months and also began providing additional officers on 12-month long tours as Technical Advisors with the various Afghan NGOs involved in the program.

Under the Demining Program, training for the Afghans was broadened to include the surveying, planning, conduct and supervision of mine clearance activities. One of the Afghan NGOs also specialised in providing mine and unexploded awareness training to refugees.

As forces were building up in Kuwait for the first Gulf War in late 1990 and early 1991, the risk to the western national contingents in Peshawar and Quetta become more acute and over the space of a few months all the national contingents except those from New Zealand and Australia ceased contributing to the program.

In early 1991, the Australian Army officer who was the Technical Advisor to Afghan Technical Consultants began travelling into Afghanistan to inspect demining operations and provide feedback into the training regime. From mid-1991 other members of the UNMCTT began to travel to inspect and expanding program of demining operations in Afghanistan. By late-1991 the New Zealanders finished their contribution and only the Australian contingent remained.

The long-term aim for the Demining Program was for it to become completely run by Afghans with no requirement for expatriate military assistance. Consequently, Afghans also became Demining Instructors in their own right and progressively took over the conduct of training. Throughout 1992 each of the Afghan NGOs in the Demining Program continued to grow in size and expertise. Increasingly Afghans took on responsibility for the planning and administration of the NGOs.

By 1993, the Demining Program had achieved a good degree of momentum and it was assessed by the Australian Government that the support from Australian contingents was no longer necessary—although several expatriate civilians continued to hold some appointments. Consequently, the last Australian contingent completed its tour of duty in mid-1993 and the last Australian technical advisors completed their tours at the end of 1993.

The Soviet-Afghan War 1979-1989

The Soviet–Afghan War lasted over nine years, from December 1979 to February 1989. Insurgent groups known collectively as the mujahideen, as well as smaller Maoist groups, fought a guerrilla war against the Soviet Army and the Democratic Republic of Afghanistan government, mostly in the rural countryside.

The mujahideen groups were backed primarily by the United States, Saudi Arabia, and Pakistan, making it a Cold War proxy war. Between 562,000 and 2,000,000 civilians were killed and millions of Afghans fled the country as refugees, mostly to Pakistan and Iran.

Prior to the arrival of Soviet troops, Afghanistan's communist party took power after a 1978 coup, installing Nur Mohammad Taraki as president. The party initiated a series of radical modernization reforms throughout the country that were deeply unpopular, particularly among the more traditional rural population and the established traditional power structures.

The government's Stalinist-like nature of vigorously suppressing opposition, executing thousands of political prisoners and ordering massacres against unarmed civilians, led to the rise of anti-government armed groups, and by April 1979 large parts of the country were in open rebellion. The government itself experienced in-party rivalry, and in September 1979 Taraki was murdered under orders of his rival and Minister of Foreign Affairs, Hafizullah Amin, which deteriorated relations with the Soviet Union.

Eventually the Soviet government, under leader Leonid Brezhnev, decided to deploy the 40th Army on December 24, 1979. Arriving in the capital Kabul, they staged a coup, killing president Amin and installing Soviet loyalist Babrak Karmal from a rival faction. The deployment had been variously called an 'invasion' (by Western media and the rebels) or a legitimate supporting intervention (by the Soviet Union and the Afghan government) on the basis of the Brezhnev Doctrine.

In January 1980, foreign ministers from 34 nations of the Islamic Conference adopted a resolution demanding 'the immediate, urgent and unconditional withdrawal of Soviet troops' from Afghanistan,[43] while the UN General Assembly passed a resolution protesting the Soviet intervention by a vote of 104 (for) –18 (against), with 18 abstentions and 12 members of the 152-nation Assembly absent or not participating in the vote. Afghan insurgents began to receive massive amounts of aid and military training in neighbouring Pakistan and China, paid for primarily by the United States and Arab monarchies in the Persian Gulf.

As documented by the National Security Archive, 'the Central Intelligence Agency (CIA) played a significant role in asserting US influence in Afghanistan by funding military operations designed to frustrate the Soviet invasion of that country. CIA covert action worked through Pakistani intelligence services to reach Afghan rebel groups.'

Soviet troops occupied the cities and main arteries of communication, while the mujahideen waged guerrilla war in small groups operating in the almost 80 percent of the country that was outside government and Soviet control, almost exclusively being the rural countryside. The Soviets used their air power to deal harshly with both rebels and civilians, levelling villages to deny safe haven to the mujahideen, destroying vital irrigation ditches, and laying millions of land mines.

The international community imposed numerous sanctions and embargoes against the Soviet Union, and the US-led a boycott of the 1980 Summer Olympics held in Moscow. The boycott and sanctions exacerbated Cold War tensions and enraged the Soviet government, which later led a revenge boycott of the 1984 Olympics held in Los Angeles.

The Soviets initially planned to secure towns and roads, stabilize the government under new leader Babrak Karmal, and withdraw within six months or a year. But they were met with fierce resistance from the guerrillas, and were stuck in a bloody war that lasted nine years. By the mid-1980s, the Soviet contingent was increased to 108,800 and fighting increased, but the military and diplomatic cost of the war to the USSR was high.

By mid-1987 the Soviet Union, now under reformist leader Mikhail Gorbachev, announced it would start withdrawing its forces after meetings with the Afghan government. The final troop withdrawal started on May 15, 1988, and ended on February 15, 1989, leaving the government forces alone in its battle against the insurgents, which continued until 1992 when the former Soviet-backed government collapsed. Due to its length, it has sometimes been referred to as the 'Soviet Union's Vietnam War' or the 'Bear Trap' by the Western media. The Soviets' failure at the war is thought to be a contributing factor to the fall of the Soviet Union.

The United Nations and Afghanistan 1980-1995

Early in 1980, the Security Council met to consider a response to the Soviet intervention, but a draft resolution condemning it was not passed, due to the negative vote of the USSR. The matter was then taken up in the General Assembly, which held an Emergency Special Session on Afghanistan over five days, from 10 to 14 January 1980. The Assembly adopted the first of a series of 'Situation in Afghanistan' resolutions, in which it deplored the armed intervention in Afghanistan, called for the withdrawal of all foreign forces, asked States to contribute humanitarian assistance, and asked the Secretary-General to keep it informed of developments.

Various approaches to the parties were made with a view to finding a means to end the conflict, but war continued. Its effects were devastating. During the next few years about 3 million refugees fled to Pakistan and 1.5 million to Iran. Many people were also driven from the countryside to Kabul, and in total more than half of the population was displaced. Estimates of combat fatalities range between 700,000 and 1.3 million people. With the school system largely destroyed, industrialization severely restricted and large irrigation projects badly damaged, the economy of the country was crippled.

The General Assembly maintained its focus on Afghanistan throughout the 1980s, adopting a series of resolutions which called for an end to the conflict, withdrawal of foreign troops, UN assistance to find a political settlement and international help for refugees and others affected by the conflict.

In 1985, the General Assembly also began a separate consideration of the human rights situation in Afghanistan. This followed receipt of the first report from a newly appointed Special Rapporteur on human rights in that country. The first in what was to become an annual resolution on human rights and fundamental freedoms in Afghanistan was adopted on 13 December 1985. In it, the Assembly expressed its profound concern about widespread disregard for human rights and large-scale violations. It also expressed concern at the severe consequences for the civilian population of indiscriminate bombardments and military operations aimed primarily at villages and agricultural infrastructure.

In May 1986, Babrak Karmal was replaced as PDPA leader by Mohammad Najibullah, who subsequently became President in November 1987. Following the exercise of the UN Secretary-General's good offices, Afghanistan, Pakistan, the USSR and the United States signed Agreements on the Settlement of the Situation Relating to Afghanistan under UN auspices on 14 April

1988. These provided for an end to foreign intervention in Afghanistan, and the USSR began withdrawing its forces.

With the Security Council's agreement on 25 April 1988 (and subsequently authorized in resolution 622 of 31 October 1988), Secretary-General Javier Perez de Cuellar set up a mission to monitor the withdrawal of foreign forces—the UN Good Offices Mission in Afghanistan and Pakistan (UNGOMAP)—and made plans to support the anticipated repatriation of refugees. The Soviet withdrawal was completed in February 1989. The mujahideen, however, who had not signed the agreements, maintained their fight against Najibullah's government and the civil war continued.

Following the May 1987 agreement, the UN had begun strenuous efforts to coordinate humanitarian assistance. Afghanistan had long been designated by the UN as one of the world's least developed countries and war only made it more difficult to respond to the challenge of reconstruction and development. The UN Food and Agriculture Organization (FAO) estimated that the area under agricultural cultivation in Afghanistan fell by 40 per cent between 1979 and 1991.

In 1988, under the guidance of the Secretary-General's newly-appointed Coordinator for UN Humanitarian and Economic Assistance Programmes Relating to Afghanistan, a plan of action was developed jointly by UN agencies and programmes, including the UN Children's Fund (UNICEF), the UN Development Programme (UNDP), the UN High Commissioner for Refugees (UNHCR) and the World Food Programme (WFP).

The Office for the Coordination of UN Humanitarian and Economic Assistance Programmes Relating to Afghanistan sometimes used the shorter title of UN Office for the Coordinator Afghanistan and used the acronym UNOCA. The UNOCA-led humanitarian efforts became known as 'Operation Salam' (meaning 'peace').

In 1991 Operation Salam was taken over by the Secretary-General's Personal Representative at the time, Benon Sevan. In that year, WFP provided 60,000 metric tons of food to needy Afghans, while FAO provided 6,800 tons of seed and more than half a million fruit and poplar saplings. Agricultural assistance, food aid, public and maternal health services and economic recovery programmes were initiated with resources provided to the UN by the international community. But other programmes that had been planned—to repair infrastructure, provide shelter and discourage narcotics production—had to be shelved because of insufficient funds.

As civil war between various factions continued following the Soviet withdrawal, the number of civilians fleeing the country increased steadily, making Afghanistan the world's worst refugee crisis. By 1990, there were 6.3 million civilians in exile—3.3 million in Pakistan and 3 million in Iran. In addition to setting up a voluntary repatriation project, UNHCR established more than 300 villages in Pakistan for the mainly ethnic Pashtun refugees. In Iran, the mostly ethnic Tajiks, Uzbeks and Hazaras lived and found work in local communities.

In 1992, fighting intensified, making the aid effort more difficult. Rebel forces closed in on Kabul and the Najibullah government fell. On 24 April 1992, leaders of the mujahideen forces except one (Gulbuddin Hekmatyar) agreed to form a government under Sibghatullah Mojaddedi. According to the agreement, Mojaddedi would head a Transitional Council for two months. He would then be replaced by a Leadership Council—to last four months—that would be headed by Burhannudin Rabbani.

Rabbani was declared President of the Islamic State in Afghanistan in July 1992. According to the agreement (called the Peshawar Accord) he should have relinquished power in October, but didn't. By that time, Massoud, Rabbani's Defence Minister, and Hekmatyar were engaged in armed confrontation in Kabul—which had largely been spared during the Soviet occupation.

The General Assembly's annual assessment of the situation—summarized in a resolution on emergency international assistance for the reconstruction of Afghanistan noted that establishment of the Islamic State provided a new opportunity for reconstruction, welcomed the Secretary-General's efforts to draw attention to mobilizing assistance for rehabilitation and reconstruction, and sought funds for an emergency trust fund to support that rehabilitation.

In 1993, two peace accords were negotiated between President Rabbani and eight other Afghan leaders—in Islamabad on 7 March and in Jalalabad on 18 May. In these accords, the leaders agreed to form a government for 18 months, to set in motion an electoral process, to formulate a constitution, and to establish a defence council to set up a national army. In his annual report issued in September 1993, the Secretary-General observed that although the accords were encouraging, they had neither resolved the problems of the government nor removed the threat of renewed fighting around Kabul.

In December 1993, at the request of the General Assembly, the Secretary-General established the UN Special Mission to Afghanistan (UNSMA) to canvass a broad spectrum of Afghan leaders and solicit their views on how the UN could best help with national reconciliation and reconstruction. Meanwhile, the movement of civilians mirrored the ebb and flow of battlefield realities, with many refugees returning to peaceful parts of the country. In 1992, more than 1.2 million refugees returned home from Pakistan.

However, despite all these positive developments, Kabul was soon besieged again: first by various mujahideen factions, and then by the Taliban—a movement with its foundations in Kandahar. The Taliban were mostly sons and orphans of mujahideen, who had been raised in refugee camps in Pakistan and were opposed to what they saw as the corruption of the mujahideen. This round of fighting led once more to the displacement of populations, with some 350,000 people fleeing the Kabul region for camps near Jalalabad, bringing the total of internally displaced people dependent on the UN for food and sustenance to 800,000. By 1994, there were an additional 700,000 Afghan refugees, living mostly in camps in Pakistan and Iran.

In 1994, the first of a series of annual consolidated appeals to aid Afghanistan was launched. The appeals detailed the emergency needs of Afghan people and asked for funds to enable non-governmental and UN agencies to address those needs. This first appeal had some success, with donors supplying 75 per cent of the funds requested. Rehabilitation projects focussed on human development and poverty alleviation in rural communities.

From 1995, however, the annual consolidated appeals were less successful in raising the necessary funds. The 1995-1996 appeal, for example, raised only 50 per cent of the amount deemed urgent—of which practically nothing was available for crucial infrastructure repairs. However, the absence of conflict in some parts of the country made it possible to reopen some roads, allowing greater aid distribution by the UN and aid agencies.

From January to June 1995, WFP distributed more than 53,000 tons of food aid, while the UN Centre for Human Settlements helped some 10,000 families rebuild their homes. During a health campaign in 1995, nearly 2.4 million children under five years of age were immunized against polio and more than 80,000 under two years old were inoculated against measles.

Demining

Demining or mine clearance is the process of removing land mines from an area. In military operations, the object is to rapidly clear a path through a minefield, and this is often done with devices such as mine plows and blast waves. By contrast, the goal of humanitarian demining is to remove all of the landmines to a given depth and make the land safe for human use. Specially trained dogs are also used to narrow down the search and verify that an area is cleared. Mechanical devices such as flails and excavators are sometimes used to clear mines.

A great variety of methods for detecting landmines have been studied. These include electro-magnetic methods, one of which (ground penetrating radar) has been employed in tandem with metal detectors. Acoustic methods can sense the cavity created by mine casings. Sensors have been developed to detect vapor leaking from landmines. Animals such as rats and mongooses can safely move over a minefield and detect mines, and animals can also be used to screen air samples over potential minefields. Bees, plants and bacteria are also potentially useful. Explosives in landmines can also be detected directly using nuclear quadrupole resonance and neutron probes.

Detection and removal of landmines is a dangerous activity, and personal protective equipment does not protect against all types of landmine. Once found, mines are generally defused or blown up with more explosives, but it is possible to destroy them with certain chemicals or extreme heat without making them explode.

Chapter Two
Selection

Once the Australian Government made the commitment to send Australian soldiers to contribute to the mine clearance training operation it then became necessary to identify to best candidates for the job. From the Army's perspective members of the Royal Australian Engineers were most suited and individuals who had a good level of training and experience in mine warfare and handling un-exploded ordnance (UXO) were the most sensible choice. However, all the people in this category were already doing other jobs throughout the Army. And while it had been months in the deliberations once the decision had been made some individuals got only days' notice that they were required to deploy. And in the Army, you were ordered to deploy; that is the nature of the job.

'Field Engineering Wing of the School of Military Engineering had been training the Australian contingent to the United Nations supervised elections in Namibia. The key risk to the contingent was ascertained to be mines. Namibia had been heavily mined by both the South African Army and the guerrilla insurgents from Angola. Field Engineering wing threw itself into the task of training up the contingent's mainly construction sappers with mine warfare skills. Many of us field engineers were disappointed that we were not able to go, however it was primarily a construction engineering role. It was also the largest

Australian Army deployment since Rhodesia a decade earlier. There were many long faces in FE wing in the final days prior to the deployment of the Namibian contingent.

'Morning tea was our most important meal at the SME Officers Mess. One would miss breakfast to avoid vomiting it all up at PT in the early morning. So, by the time morning tea came around at 10 am we were normally starving. I recall reaching for my third toasted ham and cheese sandwich only to be delayed by the School Training Officer (Major Ken Gillespie). He asked if I would be interested in leading a UN contingent to train Afghan refugees in mine warfare. I couldn't believe my luck.'

Preparations were to be kept secret as the government had not officially committed to providing the troops and there had been no public announcements. I was told I could select whoever I wanted from the entire Corps of Royal Australian Engineers... and what a team we put together. All were top rated field engineers. All had extensive experience in mine warfare training as well as running instructional courses. We had two senior warrant officers who had mine warfare experience in Vietnam (WO1 'Monty' Avotins, and WO1 'Arnie' Palmer). I had served with Arnie in Rhodesia where he was awarded an MBE for his outstanding work. Monte was a field engineering legend. Warrant Officers 'Tony' Smith and 'Jock' Turner were very experienced. Staff Sergeants' 'Shorty' Coleman and 'Bob' Kudyba came highly recommended. I was supported by two very capable Captains—Karl Chirgwin as the contingent 2IC and Paul Petersen.'

Graham Costello, 1st Contingent

'I just got back into Brisbane from an exercise when I got the word that I had to fly out within a week. I had to get back to Sydney where I was posted. Telling my wife that I was off for somewhere between 3 and 12 months and leaving her with our young child certainly didn't put me in the good books—but of course, she was understanding and supportive as that was part and parcel of Army life.'

Carl Chirgwin, 1st Contingent

'My unit was on exercise up at Pine Creek in the Northern Territory. The local copper passed me a message to ring the personnel guys at Army Headquarters. I was told that I was being sent to Peshawar. I drove to Darwin and then flew back to Melbourne. Along the way I found an atlas and looked up where Peshawar actually was.'

Paul Petersen, 1st Contingent

'At the time Australia was also getting ready to deploy a contingent to Namibia, the United Nations Transitional Assistance Group (UNTAG). At this time, I was serving in Brisbane and having not been included in the first UNTAG contingent, I suddenly found myself included in UNMCTT. What followed was a whirlwind time of moving from Brisbane to Canberra, to Sydney and then on to Pakistan.'

Bob Kudyba, 1st Contingent

'The value we placed on an opportunity like a UNMCTT deployment in 1989 needs to be understood in the context of 'the long peace' that followed Australia's commitment to Vietnam and the other warlike adventures that had occupied its armed forces since the late 1940s. Those of us who had joined after 1972 served in a peacetime Army; operational service held a special mystique and anyone who had seen it was held in high regard. As with all young service people, we itched to be tested in an operational environment and—let's admit it—to earn the cherished medal ribbons that came with that experience. Those opportunities were scarce until the late '80s. Things that approximated operational service, such as one of the long-running peacekeeping missions in the Middle East or Kashmir, or a tour with the Rifle Company at Butterworth in Malaysia, were highly sought after, but the numbers dictated that few people—and especially few Sappers—got those chances.

Things began to change in the late '80s, when developments in the international security environment and in Australia's foreign policy saw a new willingness to employ the military— by then re-branded as the Australian Defence Force—beyond our shores. This started in 1988 when Australia contributed, at quite short notice, a few observers to oversee the ceasefire at the end of the Iran-Iraq war. As this included a couple of my classmates, like many of my Army contemporaries I began to worry that I had missed a fleeting opportunity. This continued into 1989, when Australia's long-dormant commitment to support the UN Transition Assistance Group in Namibia suddenly flowered into the first unit-level Army deployment since Vietnam. As this was an Engineer-heavy deployment, my pangs of 'deployment envy' increased. By that time, I had just started a new job as the Aide-de-Camp (ADC) to the Commander of the 1st Division in Brisbane, from which I thought it very unlikely that I would be released for an operational tour. Later that year, I became peripherally aware that Australia was preparing another significant deployment to Cambodia: clearly, my hat was not in the ring for that trip. Then one day I opened a copy of the fortnightly Army newspaper to see that a small contingent of Sappers, including some close mates, had deployed to Pakistan to teach mine clearance techniques to Afghan refugees—I don't think the term 'UNMCTT' even appeared in the article, or that it would have meant anything if it did. I began to resign myself to having missed the deployment bus.

By September 1989 I'd forgotten about the UNMCTT. Then one day I received a phone call from the Engineer Career Advisor at the (then) Office of the Military Secretary, which was the agency responsible for Army officer postings. He advised that I was under consideration for the second Australian contingent of the UNMCTT in a place called Peshawar in Pakistan, to rotate in behind the 1st contingent during October. He said that this depended on my boss releasing me for deployment: that question would be put to him shortly.

My boss at the time was Major General Arthur Fittock, a distinguished and popular senior Sapper for whom I had then been working happily for about nine months. Officially, ADC appointments were for 12 months, so I didn't rate my chances of release very highly. As the ADC, I was aware of all of his phone calls so I knew when he'd received the one from the Military Secretary. I was on tenterhooks when, shortly afterwards, I had to take something into his office. Mercifully, the boss came to the point quickly: he said he'd been asked to release me and that, philosophically, he felt must do so: after all, operations were the Army's core business. I could go. I think I floated out of his office.'

Andrew Smith, 2nd Contingent

'My involvement with the UNMCTT began on 11 September 1989 when as a Staff Sergeant in 18 Field Squadron in Townsville. I received a call from then Warrant Officer Ted Bell inquiring as to whether I was interested in accepting a deployment to Pakistan. I informed Ted that I would love to accept, but my wife was due to give birth at any time. Ted replied that I should get back to him by 0900hr the next day with a 'yes' or 'no'. I went home and explained the situation to my wife that afternoon. At 0620hr on 12 September 1989 we welcomed our baby son into the world. It was also our wedding anniversary. My wife gave me the thumbs up at around 0730hr and I contacted Ted prior to 0900hr as instructed and accepted the deployment.'

Allan Mansell, 2nd Contingent

'I had been selected to be a military observer for the UN mission in Iran and Iraq but a few weeks before I was due to depart peace broke out and the mission got cancelled. Then Bill Van Ree wasn't able to return to Pakistan so they asked me about going over to be the Technical Adviser to Afghan Technical Consultants for 12 months. After chatting with Bill about what might be involved. I learned that whoever was selected for the job has already been given permission to travel across the border into Afghanistan in order to monitor the demining operations. He also told me that because the posting was for twelve months that my wife could join me— even though she was five months pregnant. I said yes!'

Graeme Membrey, Technical Advisor

As Australian support to the demining program continued the degree of notice given to the participants for each contingent improved but given the vagaries of the personnel management system invariably there were several who still received relatively short notice to deploy.

> 'I was posted to the Directorate of Engineers. It was an interesting time in the Corps with our commitment to Namibia winding down and our commitment to Peshawar well under way. Peter Kube had returned from the 3rd Contingent to Peshawar and joined the Corps Director's staff and told me about the work his team had done to help the Afghan people clear up the mess of mines and unexploded ordnance that had been left behind following 10 years of Russian occupation. It sounded like a fantastic opportunity so when I heard they were after nominations for the next contingent, I chucked my name into the ring. As with most things in the Army, it went pretty quiet after that and I didn't hear anything for a while until one day in October, then Lieutenant Colonel Ken Gillespie stuck his head in the door and said 'Kav, you're off to Pakistan. Congratulations!' and walked off.'

Michael Kavanagh, 5th Contingent

Captain Marcus Fielding on exercise in 1991.

'I remember being told that I was going to deploy in late November 1991. The challenge was that between then and deployment in February my wife and I had to move from Melbourne to Brisbane where I had been posted to the 2ⁿᵈ Combat Engineer Regiment. Then as we were moving, we found out that my wife was pregnant with our first child. We didn't have much time to settle in in Brisbane before I had to head down to Sydney for my pre-deployment training. Our son was born in September about two weeks after I got back to Australia. I have no idea what my wife's pregnancy was like...'

Marcus Fielding, 8th Contingent

'Extreme disappointment is an understatement after Darrell Crichton and I had been removed from the 1st Australian Contingent going to UNTAG in Namibia in 1989. Darrell and I were Section Commanders with the Field Engineer Troop selected for the deployment. After training our Sections, we were removed four weeks prior to the Contingent's departure due to panelling on the Subject 4 Sergeant Field Engineer promotion course. Due to our follow-on postings, we would miss the UNTAG rotations completely therefore when the UNMCTT commenced, our next opportunity to test our mettle arrived. We were elated when Darrell deployed on the 8th Contingent, and soon after completing my Explosive Ordnance Disposal (EOD) course I was warned for deployment with 9th Contingent. Each UNMCTT contingent consisted of mine warfare qualified and EOD qualified personnel and in our contingent a medic due to the increased risk levels. The mix complimented the mission due to the combined land mine and explosive remnants of war hazards. The rank break-up provided the right balance between leadership, planner, and practitioner.'

Craig Egan, 9th Contingent

Land Mines

A land mine is an explosive device concealed under or on the ground that is designed to destroy or disable enemy targets, ranging from human combatants to vehicles and tanks, as they pass over or near it. Such a device is typically detonated automatically by the pressure of a target stepping on or driving over it, although other detonation mechanisms (such as tripwires) are also sometimes used. A land mine may cause damage by direct blast effect, by fragments that are propelled by the blast, or by both.

A conventional land mine consists of a casing that contains the explosive charge. It has a firing mechanism that, when actuated, triggers a 'firing train' which leads to the initiation of the explosive charge. Land mines may also be fitted with 'anti-handling' devices intended to prevent them being neutralised by an adversary.

A wide range of land mine firing mechanisms has evolved, which are actuated by a number of stimuli including pressure, movement, sound, magnetism and vibration. Advanced anti-vehicle mines are able to sense the difference between friendly and enemy types of vehicles. This could theoretically enable friendly forces to use the mined area while denying access to the enemy. Some types of modern land mines are designed to self-destruct, or otherwise render themselves inert after a certain period. This is intended to reduce the likelihood of civilian casualties, especially post-conflict. These self-destruct mechanisms are not absolutely reliable, and most land mines that have been laid do not have these features.

Although rudimentary devices similar in function to land mines fist appeared in the mid-19th century, mines did not see widespread use until the Second World War. From that time, unexploded land mines have posed a significant humanitarian problem, causing casualties to civilians attempting to resume normal activities after conflicts and rendering large areas of land unavailable for economic activity, such as agriculture. Land mine use expanded significantly during the Cold War 'proxy wars' of the 1970s and 1980s. This coincided with the increasing use of plastic and other materials in mine designs, making them cheaper, longer-lasting and more difficult to detect and clear. By the late 1980s, land mine contamination was a major humanitarian problem in conflict-affected areas. Currently, sixty countries are contaminated with land mines, which kill more than 2,000 people every year and main more than 4,000. Most landmine casualties are civilian and children are the most affected age group. Most deaths occur after open hostilities end.

The humanitarian aid 'sector' consisting of the United Nations, Red Cross and a number of other international and non-government organisations, responded to the land mine problem by implementing organised humanitarian demining programs. By the mid-1990s, these had evolved into a de facto international demining 'industry', with emerging standards, codes of best practice, and 'doctrine.'

The terrible humanitarian consequences of land mine abuse led, by the early 1990s, to a movement to ban or severely restrict their use. This led, by the mid-1990s, to the establishment of an International Campaign to Ban Land Mines. This campaign resulted in the agreement, in 1997, of Convention on the Prohibition of the Use, Stockpiling, Production and Transfer of Anti-Personnel Mines and on their Destruction, also known as the Ottawa Treaty. To date, 164 nations have signed the treaty the notable exceptions of China, the Russian Federation, and the United States. One of the Convention's requirements is that countries in which land mines are emplaced must remove them. This assures that demining will remain on the international agenda for the foreseeable future.

Booby Traps

A booby trap is a contrivance that is intended to kill, harm, or surprise a person or animal performing an action that would normally be considered safe, such as opening a door, picking something up, or switching something on. They can also be triggered by vehicles driving along a road, as in the case of victim-operated improvised explosive devices (IEDs).

Lethal booby traps are often used in warfare, particularly guerrilla warfare. As the word trap implies, they sometimes use a form of 'bait' intended to lure the victim to carry out the action that will trigger the device. Traps can be set to act upon trespassers intruding on certain areas. Booby traps can take various forms, but the most common ones employ explosive charges initiated by a range of fuses or 'switches.' The Soviet MUV family of mine fuses, common in Afghanistan, was ideally suited to booby trap applications.

Unexploded Ordnance

Unexploded ordnance (UXO), unexploded bombs (UXBs), or explosive remnants of war (ERW) are explosive weapons (bombs, shells, grenades, land mines, naval mines, cluster munition, etc.) that did not explode when they were employed and still pose a risk of detonation, sometimes many decades after they were used or discarded. UXO does not always originate from wars; areas such as military training grounds can also hold significant numbers, even after the area has been abandoned.

UXO from World War I continue to be a hazard, with poisonous gas filled munitions still a problem. When unwanted munitions are found, they are sometimes destroyed in controlled explosions, but accidental detonation of even very old explosives also occurs, sometimes with fatal results.

Seventy-eight countries are contaminated by land mines, which kill 15,000–20,000 people every year while severely maiming countless more. Approximately 80% of casualties are civilian, with children as the most affected age group. An estimated average of 50% of deaths occurs within hours of the blast. In recent years, mines have been used increasingly as weapons of terror against local civilian populations specifically.

In addition to the obvious danger of explosion, buried UXO can cause environmental contamination. In some heavily used military training areas, munitions-related chemicals such as explosives and perchlorate (a component of pyrotechnics and rocket fuel) can enter soil and groundwater.

CHAPTER THREE
PRE-DEPLOYMENT

Administrative preparation and training for the deployment was conducted at the School of Military Engineering (SME) in Moorebank on the outskirts of Sydney. Initially only a few days in duration it became longer and more detailed with each contingent. At first, the information and advice at hand was either non-existent or just plain wrong, but as each contingent returned, they were able to add to the body of knowledge about the operation and therefore improve the pre-deployment training for those contingents that followed.

'We knew nothing, but we didn't even know that. Pre-deployment was a haze, fortunately the team was brilliant and with everyone responsible for specific aspects of preparation we got a lot done in a very short period of time.

Our greatest coup was personal kit. It is normally impossible to get the Army to diverge from standard issues of personal kit. Our normal kit at the time was post-Vietnam cotton 'greens' with a light jumper for warmth and a plastic raincoat in case it rained. Army office initially seemed quite comfortable with us heading off as if we were going to an exercise in Australia. We mentioned to Army Office that we were off to a war zone and the very first contingent. We dramatically painted a picture of the snow-capped highest mountain in Afghanistan—Mir Samir. To our everlasting discredit we took advantage of the situation and prised open the purse strings of Army Office logistics and got ourselves approval to go shopping at a 'civvie' camping store.

We walked away with top of the line Scarpa mountain climbing boots appropriate for the final ascent on Everest, air mattresses that inflated themselves by magic, and all sorts of warm gear. Army Office got even. We were authorised to travel business class to Pakistan. Despite our best efforts we were booked economy ... via Bombay ... via Air India. The whole plane smelled of curry. Some of us started the deployment already ill from food poisoning.

In Rhodesia, we had been armed with rifles and pistols. Likewise, the Namibian contingent was armed. Given that Afghanistan and the tribal lands in Pakistan are arguably much more violent and unpredictable, it seemed a sensible precaution to take some weapons. We were told that the Pakistani Army would be providing security. It was to be a peacekeeping mission. Weapons would not be necessary. We set off without any means of personal protection.

A briefing was conducted by the Joint Intelligence Organisation in Russell Offices. We were warned that it was 'TOP SECRET'. It was a high-level overview of the conflict with the Soviet Army and recent initiatives by the Mujahidin. They listed a couple of the key tribal groups and major mujahidin affiliations. It seemed comprehensive but at a very high level. There was no information about the situation on the ground or the people we were likely to be working with. At the airport on our departure, I bought a Newsweek magazine to read on the plane. I was surprised that it had exactly the same 'TOP SECRET' briefing.

All of the team were qualified in mine warfare. Many of us had just completed running the mine warfare training for the Namibia conflict. During that training we were frustrated by the lack of knowledge about the actual mines in use and how they were being deployed. Whilst we assumed that we were likely to come across mainly Soviet mines we had no information on more recent types. What little information there was, was dated, market 'SECRET', and not relevant to Afghanistan. It was hard to believe that the information was not available. We made it out mission to rectify this situation during the deployment.

We had recently had Sandy MacGregor present to the junior officer course at SME on his time as the first engineer troop commander in Vietnam. As a group we had been struck with how that small troop of sappers had 'punched well above their weight' by discovering the Cu Chi tunnels in Vietnam and developing tunnel warfare. I was also struck with how easily it was to lose a life through mis-adventure. Corporal Bowtell's death had been due to asphyxiation whilst clearing a narrow Cu Chi tunnel. Just like 3 Field Troop, we were going to be the first. We had no idea what we would have to do. All we knew was that we had to learn fast, do a good job, build expertise, and bring everyone home safe.'

Graham Costello, 1st Contingent

1st Contingent Members at Canberra in 1989. Back Row from Left: George 'Jock' Turner, Imants 'Monty' Avotins, Alan Palmer, Robert 'Bob' Kudyba, Anthony Smith. Front Row from Left: Paul Petersen, Carl Chirgwin, Craig Coleman, Graham Costello.
(AWM CANA/89/0316/02)

'I was deployed really quickly—like really quickly. I got the order on 15 July in Pine Creek in the Northern Territory. I was in Islamabad the next day and Peshawar the day after that.'

Paul Petersen, 1st Contingent

'I noticed on an atlas that Peshawar was at the same latitude as Armidale so I figured it was going to be pretty cool. I packed winter clothes and we were issued with cold weather gear in Sydney. But when I arrived in Peshawar it was 40 degrees C and stayed that way for my whole tour.'

Paul Petersen, 1st Contingent

'Something like four weeks remained between my official assignment and when our contingent deployed. After wrapping up my affairs in Brisbane I got down to the School of Military Engineering to join the rest of the contingent with about 10 days to go. In that time, we needed to be refreshed on key skills and training, receive a number of special briefings, get kitted out, and say final goodbyes to families.

Since that time, the ADF has gained a lot of experience at 'mounting' a force for operational deployment, but in 1989 it was still shaking out the cobwebs of the long peace. Fortunately for our contingent, the Quartermaster of SME who oversaw the kitting process was Captain 'Bill' Morley, a very experienced and professional officer with whom I had worked in a previous posting. Under his supervision that part of our preparation went very smoothly.

One complication arose in the fact that ours was to be the first contingent to 'winter over' in Peshawar. No-one seemed to know what that involved, except that it was expected to be very cold. With little experience of working in 'very cold' climates, the Army fell back on what it did know by issuing us with the Antarctic Block Scale of kit. In addition to gumboots and Second World War-vintage woollen uniforms, this involved a trip to the Paddy Palin outdoor store in Haymarket to be issued with (then) state of the art cold weather gear. This included a high-performance down sleeping bag and a pair of Scarpa hiking boots that stayed with me for over 30 years. Our physical preparation also involved commencing malaria prophylaxis. All of this led to the sense that we were embarking on something out of the ordinary.

Another key element of our preparation was an overnight trip down to Canberra for a series of briefings at Russell Offices, the formidable home of the Department of Defence, which about half of the contingent had never visited before. One briefing was from the (then) Joint Intelligence Organisation (JIO), which is memorable in retrospect for the total absence of information about Peshawar's role as a portal for foreign fighters making their way into Afghanistan to join the Mujahideen; and especially about a certain Saudi-Yemeni man who had been active in the town for the preceding few years. If the JIO briefing is any indication, the West was oblivious to the tectonic forces operating in the (then) North West Frontier Province and in Afghanistan that are symbolized nowadays by Al-Qaida. We had a chance to see those forces at work during our deployment.

The most significant interaction at Russell Offices was with the Directorate-General of Operations and Plans—Army, and in particular its Directorate of Operations, which in those days ran all of Army's overseas operations. The Director of Operations, Colonel Neil Turner, was a tough Vietnam veteran who impressed upon us in no uncertain terms that our mission rode on the 'ragged edge' of the Government's comfort zone for military activity and that we were not to do anything to disturb that comfort. He stressed that the Minister, Kim Beasley, had expressly forbidden crossing the border into Afghanistan and promised dire consequences for anyone who did so. We left Canberra with the impression that, despite the distance between Canberra and Peshawar and the apparently low profile of our mission, we would be watched very carefully.'

Andrew Smith, 2nd Contingent

'We were told that Australian had no real strategic interest in Afghanistan but was seeking to be seen as a 'good international citizen'. Apparently, this was primarily a humanitarian mission and NOT a military mission.'

Anonymous, 2nd Contingent

'When seated for one of the many briefings that our team received at Russell Offices, we were informed that the mission would involve us training Afghan refugees at Peshawar, Pakistan in humanitarian mine clearance. At that point Staff Sergeant Ian Mahoney sitting next to me asked me what humanitarian demining was. I replied that I didn't have a clue and that we could ask at the end of the briefing. The opportunity presented itself and I posed the question of what humanitarian demining was, to the person who was briefing us and the response was 'I don't know either, apparently you blokes are going to be involved in inventing it'. I turned back to Ian and said 'There's your answer mate.''

Allan Mansell, 2nd Contingent

'When I went to Canberra for my intelligence briefing, they asked 'Who's the silly fool being sent to Afghanistan?' I proudly replied, 'it's me!' and was led off for more detailed briefings. After receiving the briefings, I asked about carrying weapons and was told sharply, 'No'. I then asked about emergency extraction and was told 'None.' I asked about life saving support and was told 'Look Graeme, this is a high risk posting. The minister of Defence had to get the Prime Minister's concurrence. There really is no protection or support for you over there. You're basically on your own....' I realised I was going to have to survive on my common sense with some good luck.'

Graeme Membrey, Technical Advisor

'The first contingent came back and produced a training information bulletin about what they had learned and this was eagerly studied during the pre-deployment training. It included a lot of information about the types of mines that were being encountered and the emerging clearance tactics and techniques that were being developed.'

Anonymous, 3rd Contingent

'A standing Army policy which inhibited our effectiveness as a team was our inability to visit the demining sites in Afghanistan. This had been a policy since the 1st Contingent. I suspect that every contingent noted in handovers and debriefing material that it was essential to observe how well the training was being implemented.

This of course meant travelling across the border into Afghanistan and accepting whatever hazards that may have arisen. As a slightly humorous side, at the completion of a pre-embarkation briefing in Canberra I posed the question, 'Do we have permission to enter Afghanistan for the purpose of observing training?' The Lieutenant Colonel immediately responded in the affirmative and a number of the senior members of our team looked at each other and smiled. This was a coup; we were about to do something no other team had done. Happy with the response we left the building and started towards our transport. Not long after exiting the building we were stopped by shouts from behind. We turned to see the briefing officer hot footing it after us to set the record straight. 'No' we were not to enter Afghanistan. If only we had walked a little quicker!! As it turned out the 6th Contingent—our successors—were given approval.'

Brian O'Connell, 5th Contingent

'We came together as a team at SME in early November 1990. Major Brian 'Bear' O'Connell was the Officer Commanding Nuclear, Biological and Chemical Defence Wing at SME, I was from the Directorate of Engineers, Warrant Officer Dave Edwards was the Regimental Sergeant Major of SME and Sergeant Mick McQuinn was an Instructor in Field Engineering Wing so it was either a good time to be posted to Casula or they didn't cast the net very far for the 5th Contingent. Either way, it was good for team cohesion as most of us knew each other or knew of each other. That left just Corporal Dean 'Browndog' Brown from 18 Field Squadron in Townsville and Sergeant Barry Pickering, an Assault Pioneer from the School of Infantry, to round out the team. The decision was made to include an Assault Pioneer in the 4th Contingent and this was continued for our deployment.

Being the 5th Contingent, the pre-deployment training provided at SME was pretty well developed by the time we arrived. We underwent refresher training in mine warfare, unexploded ordnance, various types of fuses we were likely to encounter, explosive disposal techniques and some practical activities on the Demolition Range. My recollections are that it was very professionally delivered and I was impressed that this much effort was being invested in our team. We were also issued quite a bit of kit, not just from the SME Q-Store but we also got to go across Moorebank Avenue to the Defence National Storage and Distribution Centre. Amongst other things, we were issued extra security trunks, Alice Packs, specialist hiking boots, I don't know how many sets of long johns and sleeping bags that were rated to minus 12 degrees Celsius. We also undertook some counselling sessions with the Army Psychologists which seemed to focus on conflict resolution within teams and how we were going to get along living in a house with each other for the best part of five months, which I remember being well intentioned but didn't really hit the mark.

Part of the pre-deployment training also included two days of briefings in Canberra. We drove ourselves down in a minibus in the morning and spent the remainder of the day in briefings which I don't remember. We then went to our accommodation at the Kingston Hotel which was our first chance at 'team bonding' over dinner facilitated by quite a few beverages. The next morning was the brief on Afghanistan which had become quite famous by this stage for being of absolutely no use in preparing us for what we were going to encounter, and the brief lived up to its expectations. It was delivered by a British Exchange Officer and went into the strategic issues in Afghanistan which didn't make much sense to me at the time.

After a few more briefs it was back in the minibus and return to SME to see the medics and receive our injections. We received quite a few inoculations and it didn't seem to matter if we had received them recently or not, we were getting a booster. I remember one in particular and I believe it was called immunoglobulin which was supposed to boost our immune system to combat all the bugs we were going to be exposed to. The reason I remember it was because the syringe was huge. We received one mil for each month we were going to be deployed so they gave us five or six mils but the medic advised that it was difficult to inject so it was mixed with another 2 mils of local anaesthetic. It was injected into the buttock region and after it was administered, I remember thinking that someone had put a wallet under my skin. By the next morning, it had been absorbed and the wallet was gone.'

Michael Kavanagh, 5th Contingent

'While we were doing pre-deployment training, we were told that our deployment was going to be extended from four to six months and that we were going to be able to do visits to monitor demining sites in Afghanistan. We knew that Graeme Membrey had just started to do these visits but we really had no clear idea of what to expect. We also had a few late additions to our contingent. There was a lot of excitement, but after we all arrived in Peshawar in March no one from our contingent actually managed to get across the border until June.'

Michael Lavers, 6th Contingent

'It was interesting to be ordered to grow a beard instead of having to shave every day as per normal Army routine. The process was allowed to start when we began the pre-deployment training and so while we were there we started looking more 'rugged' every day. We were told that sporting a beard was necessary to give us more status in Pakistan and Afghanistan and to also help us to blend in.'

Marcus Fielding, 8th Contingent

'I had only been a young officer in the Corps for a few years and was pretty green. I realise now that I had no real appreciation for what for what I was about to be involved in. I could get my head around training locals in mine clearance work, but the recent change of policy that approved us going into Afghanistan was a whole new ball game. Wasn't there still a war going on there? Did I think twice about going? Hell no!'

Marcus Fielding, 8th Contingent

Soldiers love to bitch about how crappy their issued equipment is—because it usually is. And so, since this was a serious mission we were to be issued with some proper kit. But instead of going down to the Q-store we went down to Paddy Palin and came out with bags and bags of great gear. Gore-Tex jackets and gloves—almost magic. But of course, it was all formally issued to us and we expected to have to hand it all back when we returned. The Army never actually gives you anything.'

Marcus Fielding, 8th Contingent

'We concentrated as a formed body for mission specific training at SME. Dean Beaumont had returned from the 8th Contingent and took the lead on our preparations. Dean took it on himself to ensure our training was adapted in accordance with the changing mission. He ran basic Pashtu language and cultural training, briefed us on security aspects, and refreshed our mine awareness and clearance techniques. Noting in the early 1990s the extent of simulation training consisted of practice mines with practice fuses that would provide a 'pop' and smoke effect if initiated. Understanding these limitations, and to increase the sense of realism in our preparations, a proposal was submitted to have us train with fused and armed mines. The idea being, if we overcome the psychological pressures of conducting live mine clearance prior to deployment, it would raise our confidence and increase our resilience to what may await our deployment. Dean had requested approval to arm the mines however the risk was deemed too high. We completed our training with live fused mines, but the pins remained in place. The blend of cultural, language, situational awareness, and intense knowledge skills and attitude development in our pre-deployment training was exceptional. When combined with the exceptional kit and clothing provided by Mick Collins and the logistic team, we were prepared effectively for the mission. A great job by all involved.'

Craig Egan, 9th Contingent

Pakistan

Pakistan, officially the Islamic Republic of Pakistan, is a country in South Asia. It is the fifth-most populous country with a population exceeding 212,742,631 people. In area, it is the 33rd-largest country, spanning 881,913 square kilometres. Pakistan has a 1,046-kilometre coastline along the Arabian Sea and Gulf of Oman in the south and is bordered by India to the east, Afghanistan to the west, Iran to the southwest, and China in the far northeast. It is separated narrowly from Tajikistan by Afghanistan's Wakhan Corridor in the northwest, and also shares a maritime border with Oman.

The territory that now constitutes Pakistan was the site of several ancient cultures and was later home to kingdoms ruled by people of different faiths and cultures, including Hindus, Indo-Greeks, Muslims, Turco-Mongols, Afghans, and Sikhs. The area has been ruled by numerous empires and dynasties, including the Persian Achaemenid Empire, Alexander III of Macedon, the Indian Mauryan Empire, the Arab Umayyad Caliphate, the Gupta Empire, the Delhi Sultanate, the Mongol Empire, the Mughal Empire, the Afghan Durrani Empire, the Sikh Empire (partially), and, most recently, the British Empire.

Pakistan is the only country to have been created in the name of Islam. As a result of the Pakistan Movement led by Muhammad Ali Jinnah and the subcontinent's struggle for independence, Pakistan was created in 1947 as an independent homeland for Indian Muslims. It is an ethnically and linguistically diverse country, with a similarly diverse geography and wildlife.

Initially a dominion, Pakistan adopted a constitution in 1956, becoming an Islamic republic. An ethnic civil war and Indian military intervention in 1971 resulted in the secession of East Pakistan as the new country of Bangladesh. In 1973, Pakistan adopted a new constitution which stipulated that all laws are to conform to the injunctions of Islam as laid down in the Quran and Sunnah.

A regional and middle power, Pakistan has the sixth-largest standing armed forces in the world and is also a nuclear power as well as a declared nuclear-weapons state, the second in South Asia and the only nation in the Muslim world to have that status.

Pakistan has a semi-industrialised economy with a well-integrated agriculture sector and a growing services sector. It is ranked among the emerging and growth-leading economies of the world, and is backed by one of the world's largest and fastest-growing middle class.

Pakistan's political history since independence has been characterized by periods of military rule, political instability and conflicts with India. The country continues to face challenging problems, including overpopulation, terrorism, poverty, illiteracy, and corruption.

Peshawar

Peshawar is the capital of the Pakistani province of Khyber Pakhtunkhwa (North West Frontier Province at time of UNMCTT). Up until 2010 the province was called the North-West Frontier Province. Peshawar also serves as the administrative centre and economic hub for the Federally Administered Tribal Areas (FATA).

Peshawar is located 260 km west of Islamabad.

Situated in a broad valley near the eastern end of the historic Khyber Pass, about 60 km from the border with Afghanistan, Peshawar's recorded history dates back to at least 539 BCE, making it the oldest city in Pakistan and one of the oldest in South Asia.

The city was an important trading centre during the Mughal era before serving as the winter capital of the Afghan Durrani Empire from 1757 until the city was captured by the Sikhs in 1818, who were then followed by the British in 1849.

Following the defeat of the Sikhs in the Second Anglo-Sikh War in 1849, territories in the Punjab were also captured by the British East India Company. The British for re-established stability in the wake of ruinous Sikh rule. During the Sepoy Rebellion of 1857, the 4,000 members of the native garrison were disarmed without bloodshed; the absence of brutality meant that Peshawar was not affected by the widespread devastation that was experienced throughout the rest of British India and local chieftains sided with the British after the incident.

The British laid out the vast Peshawar Cantonment to the west of the city in 1868, and made the city its frontier headquarters. Additionally, several projects were initiated in Peshawar, including linkage of the city by railway to the rest of British India and renovation of the Mohabbat Khan mosque that had been desecrated by the Sikhs. British suzerainty over regions west of Peshawar was cemented in 1893 by Sir Mortimer Durand, foreign secretary of the British Indian government, who collaboratively demarcated the border between British controlled territories in India and Afghanistan.

For better administration of the region, Peshawar and the adjoining districts were separated from the Punjab Province in 1901, after which Peshawar became capital of the new province.

In 1947, Peshawar became part of the newly created state of Pakistan, and emerged as a cultural centre in the country's northwest. The University of Peshawar was established in the city in 1950, and augmented by the amalgamation of nearby British-era institutions into the university. Until the mid-1950s, Peshawar was enclosed within a city wall and sixteen gates.

In the 1960s, Peshawar was a base for a CIA operation to spy on the Soviet Union, with the 1960 U-2 incident resulting aircraft an aircraft flown from Peshawar was shot down by the Soviets. From the 1960s until the late 1970s, Peshawar was a major stop on the famous 'Hippie Trail'.

Peshawar is overwhelmingly Muslim, with Muslims making up 98.5% of the city's population. Peshawar's economic importance has historically been linked to its privileged position at the entrance to the Khyber Pass— the ancient travel route by which most trade between Central Asia and the Indian Subcontinent passed. Peshawar's economy also benefited from tourism in the mid-20th century, as the city formed a crucial part of the Hippie trail.

Peshawar has hosted Afghan refugees since the start of the Afghan civil war in 1978, though the rate of migration drastically increased following the Soviet invasion of Afghanistan in 1979. By 1980, 100,000 refugees a month were entering the province, with 25% of all refugees living in Peshawar district in 1981. The arrival of large numbers of Afghan refugees strained Peshawar's infrastructure, and drastically altered the city's demography. During the 1988 national elections, an estimated 100,000 Afghans refugees were illegally registered to vote in Peshawar.

During the Soviet war in Afghanistan in the 1980s, Peshawar served as a political centre for the CIA and the Inter-Services Intelligence-trained mujahideen groups.

With an influence from the local steppe climate, Peshawar features a hot semi-arid climate, with hot summers and cool winters. Winter in Peshawar starts in November and ends in late March, though it sometimes extends into mid-April, while the summer months are from mid-May to mid-September. The mean maximum summer temperature surpasses 40°C during the hottest month, and the mean minimum temperature is 25°C. The mean minimum temperature during the coolest month is 4°C, while the maximum is 18.3°C. Peshawar is not a monsoon region, unlike other parts of Pakistan; however, rainfall occurs in both winter and summer.

The Federally Administered Tribal Areas

The Federally Administered Tribal Areas (FATA) was a semi-autonomous tribal region in north-western Pakistan that existed from 1947 until being merged with neighbouring province Khyber Pakhtunkhwa in 2018.

The FATA consisted of seven tribal agencies (districts) and six frontier regions, and were directly governed by Pakistan's federal government through a special set of laws called the Frontier Crimes Regulations. It bordered Pakistan's provinces of Khyber Pakhtunkhwa and Balochistan to the east and south, and Afghanistan's provinces of Kunar, Nangarhar, Paktia, Khost and Paktika to the west and north. The territory is almost exclusively inhabited by the Pashtun, who also live in the neighbouring provinces of Khyber Pakhtunkhwa and Northern Balochistan, and straddle across the border into Afghanistan. They are mostly Muslim.

Since the 9/11 attacks in the United States in 2001, the tribal areas are a major theatre of militancy and terrorism. Pakistan Army launched 10 operations against the Taliban since 2001, most recently Operation Zarb-e-Azb in North Waziristan. The operations have displaced about two million people from the tribal areas, as schools, hospitals, and homes have been destroyed in the war.

On 2 March 2017, the federal government considered a proposal to merge the tribal areas with Khyber Pakhtunkhwa, and to repeal the Frontier Crimes Regulations. However, some political parties have opposed the merger, and called for the tribal areas to instead become a separate province of Pakistan.

On 24 May 2018, the National Assembly of Pakistan voted in favour of an amendment to the Constitution of Pakistan for the FATA-KP merger which was approved by the Senate the following day. Since the change was to affect the province of Khyber Pakhtunkhwa, it was presented for approval in the Khyber Pakhtunkhwa Assembly on 27 May 2018, and passed with majority vote. On 28 May 2018, the President of Pakistan signed the FATA Interim Governance Regulation, a set of interim rules for FATA until it merges with Khyber Pakhtunkhwa within a timeframe of two years.

The 25th Amendment received assent from President Mamnoon Hussain on 31 May 2018, after which FATA was officially merged with Khyber Pakhtunkhwa. This further weakened the Pashtunistan movement in a historical context, as Pakistan's government established full rule, including legal system over the territory.

The Durand Line

The Durand Line is the 2,430-kilometre international border between Pakistan and Afghanistan. It was established in 1896 between Sir Mortimer Durand, a British diplomat and civil servant of the British Raj, and Abdur Rahman Khan, the Afghan Amir, to fix the limit of their respective spheres of influence and improve diplomatic relations and trade.

Afghanistan was considered by the British as an independent state at the time, although the British controlled its foreign affairs and diplomatic relations.

The single-page agreement, dated 12 November 1893, contains seven short articles, including a commitment not to exercise interference beyond the Durand Line. A joint British-Afghan demarcation survey took place starting from 1894, covering some 800 miles of the border.

Established towards the close of the 'Great Game', the resulting line established Afghanistan as a buffer zone between British and Russian interests in the region. The line, as slightly modified by the Anglo-Afghan Treaty of 1919, was inherited by Pakistan in 1947 following its independence.

The Durand Line cuts through the Pashtun tribal areas and further south through the Balochistan region, politically dividing ethnic Pashtuns, as well as the Baloch and other ethnic groups, who live on both sides of the border.

It demarcates Khyber Pakhtunkhwa, the Federally Administered Tribal Areas, Balochistan and Gilgit-Baltistan of northern and western Pakistan from the north-eastern and southern provinces of Afghanistan. From a geopolitical and geostrategic perspective, it has been described as one of the most dangerous borders in the world.

Although the Durand Line is recognized as the western border of Pakistan, it remains largely unrecognized by Afghanistan. In 2017, amid cross-border tensions, former Afghan President Hamid Karzai said that Afghanistan will 'never recognise' the Durand Line as the international border between the two countries.

The Khyber Pass

The Khyber Pass is a mountain pass in the northwest of Pakistan, on the border with Afghanistan. It connects the town of Landi Kotal to the Valley of Peshawar at Jamrud by traversing part of the Spin Ghar mountains. The summit of the pass is 5 km inside Pakistan at Landi Kotal, while the lowest point is at Jamrud in the Valley of Peshawar.

An integral part of the ancient Silk Road, it has long had substantial cultural, economic, and geopolitical significance for Eurasian trade. Throughout history, it has been an important trade route between Central Asia and South Asia and a vital strategic military choke point for various states that came to control it.

Well known invasions of the area have been predominantly through the Khyber Pass, such as the invasions by Darius I, Genghis Khan and later Mongols such as Duwa, Qutlugh Khwaja and Kebek.

Prior to the Kushan era, the Khyber Pass was not a widely used trade route. The Khyber Pass became a critical part of the Silk Road, which connected Shanghai in the East to Cádiz on the coast of Spain. The Parthian and Roman Empires fought for control of passes such as these to gain access to the silk, jade, rhubarb, and other luxuries moving from China to Western Asia and Europe.

Through the Khyber Pass, Gandhara (in present-day Pakistan) became a regional center of trade connecting Bagram in Afghanistan to Taxila in Pakistan, adding Indian luxury goods such as ivory, pepper, and textiles to the Silk Road commerce.

Among the Muslim invasions of the Indian subcontinent, the famous invaders coming through the Khyber Pass are Mahmud Ghaznavi, and the Afghan Muhammad Ghori and the Turkic-Mongols. Finally, Sikhs under Ranjit Singh captured the Khyber Pass in 1834 until they were defeated by the forces of Wazir Akbar Khan in 1837.

To the north of the Khyber Pass lies the country of the Mullagori tribe. To the south is Afridi Tirah, while the inhabitants of villages in the Pass itself are Afridi clansmen. Throughout the centuries the Pashtun clans, particularly the Afridis and the Afghan Shinwaris, have regarded the Pass as their own preserve and have levied a toll on travellers for safe conduct. Since this has long been their main source of income, resistance to challenges to the Shinwaris' authority has often been fierce.

For strategic reasons, after the First World War the British built a heavily engineered railway through the Pass. The Khyber Pass Railway from Jamrud, near Peshawar, to the Afghan border near Landi Kotal was opened in 1925. During World War II concrete 'dragon's teeth' (tank obstacles) were erected on the valley floor due to British fears of a German tank invasion of British India.

The Pass became widely known to thousands of Westerners and Japanese who travelled it in the days of the 'Hippie Trail', taking a bus or car from Kabul to the Afghan border. At the Pakistani frontier post, travellers were advised not to wander away from the road, as the location was a barely controlled Federally Administered Tribal Area (FATA). Then, after customs formalities, a quick daylight drive through the Pass was made. Monuments left by British and Indian Army units, as well as hillside forts, can be viewed from the highway.

The area of the Khyber Pass has been connected with a counterfeit arms industry, making various types of weapons known to gun collectors as Khyber Pass copies, using local steel and blacksmiths' forges.

During the war in Afghanistan, the Khyber Pass has been a major route for resupplying military armament and food to the NATO forces in the Afghan theatre of conflict since the US started the invasion of Afghanistan in 2001. Almost 80 percent of the NATO and US supplies that are brought in by road were transported through the Khyber Pass. It has also been used to transport civilians from the Afghan side to the Pakistani one.

Until the end of 2007, the route had been relatively safe since the tribes living there (mainly Afridi, a Pashtun tribe) were paid by the Pakistani government to keep the area safe. However, after that year, the Taliban began to control the region, and so there started to exist wider tensions in their political relationship.

Since 2008, supply convoys and depots in this western part have increasingly come under attack by elements from or supposedly sympathetic to the Pakistani Taliban.

CHAPTER FOUR
ARRIVAL AND ORIENTATION

The first contingent flew into Islamabad and then travelled by bus to Peshawar. Later contingents usually flew into Karachi and then transferred onto a domestic flight to Peshawar. The first contingent was met by the Defence Advisor at the Australian Embassy. Subsequent contingents were met at Peshawar by members of the outgoing contingent. Knowing that the arriving contingents would be 'amused' by the environment it became traditional for the outgoing contingent to make sure that the incoming contingent got the full local experience on arrival in Peshawar. All of the contingents lived in a new suburb of Peshawar called Hayatabad.

> 'The UN had rented a large concrete house in a new development on the outskirts of Peshawar. There were lots of empty housing blocks with the occasional large concrete house. All houses were surrounded by a high concrete wall with iron gates and guarded by a chowkidar. Our chowkidar did not fill us with confidence. He slept most of the day and all the night on a mat in his guard room. He must have been one hundred years old, weighed 50 kg, and never looked well. He was not armed. With the benefit of hindsight, the building had been perfectly positioned for a car bombing or at least RPG practice. Our only consolation was that if we were attacked the chowkidar's tribe would take it as a personal offence and set about retaliation.

No doubt at least one of our staff was Pakistani security service and we were constantly under close supervision and loose cover protection.

Neighbours did not invite us to visit. Indeed, the first sign of neighbour discontent was a few rounds fired in our general direction. Naively we were going for runs around the neighbourhood trying to keep up our fitness. A few AK-47 rounds in the near vicinity was enough to curtail that activity. Likewise, we had extorted from Army Office some funds to buy some basic gym equipment that we set up on the roof. Unfortunately, our presence in shorts and t-shirts on our roof raised a neighbourly complaint—the standard format of a couple of AK-47 rounds either badly aimed or nonchalantly directed. When we first went up on the roof, we were surprised to find so many expended bullets. After a while we worked out that these were just the bullets falling from the sky from recent celebrations. Everyone had a gun—except us. Some would just shoot at things for practice. One day we found a bullet lodged in the car just behind where the passenger's head would be. We expect it was probably just someone having some fun.

All the contingent members had previous experience designing and running well constructed courses. We were surprised to find that the Risalpur training ground had nothing. There was a small contingent of Pakistani Army Engineers, but they didn't seem to be that interested in formal design of training. The team set about developing detailed training design. We were being hampered by a lack of office support, so we set about the difficult task of extorting even more funds from Army Office to buy a computer. Army Office was right to be concerned as we eventually bought possibly the most expensive IBM PC in the known world. From memory it cost US$20k for a basic IBM PC with 3½ inch floppy disk drive, green screen, and Word Perfect. At the time our annual salary was not much more than that.'

Graham Costello, 1st Contingent

'When we arrived in Islamabad, we were met by Colonel Brian Cloughley, the Defence Advisor at the Australian Embassy. He sported a fulsome waxed moustache and came across as very 'chipper'. He ushered us into a minivan and then broke open a case of cold beer to welcome us to Pakistan.'

Carl Chirgwin, 1st Contingent

'The day after we arrived a bicycle bomb went off in the old bazaar and killed 14 people. I distinctly remember the intelligence spooks in Australia telling us that they had not heard of any incidents for several months. When I spoke to the local police man he laughed and said they had about two or three violent incidents every day!'

Carl Chirgwin, 1st Contingent

'We stayed at the Pearl Continental Hotel for a little while and then moved into a rented house out in Hayatabad where the other contingents also had houses. Our house warming party lasted 24 hours! But on the plus side we raffled off a slouch hat and managed to raise US$300 for the Toorak Home for Wayward Boys.'

Carl Chirgwin, 1st Contingent

'We arrived at the same time as the Canadians and they really suffered in the heat. The Kiwis had been there for a while and they were terrific in settling us all in.'

Paul Petersen, 1st Contingent

'The first night we were in Peshawar there was a bomb blast that we later learned had been hidden in a bicycle in the market. Apparently, it was the fourth blast that month. I reckon there were probably a hundred blasts in the four months of my tour.'

Paul Petersen, 1st Contingent

'The house we moved into was really expensive to rent. But getting a phone connection was really expensive so we didn't or couldn't get one. Only the yanks had one but they weren't going to allow us to use it.'

Paul Petersen, 1st Contingent

Australia House at Hayatabad, Peshawar in 1992.

'Much of that time still remains a 'blur' as we shuttled from here to there to somewhere else. My memories of travelling in a rickety old bus along the Grand Trunk Road from Islamabad to Peshawar remain, the Grand Trunk Road was a seething mass of confused humanity, all travelling in an organised sort of mayhem in both directions, I still reckon it was a miracle we made it to Peshawar.

Once we arrived, we were shuttled around to do a number of administrative and procedural duties and receive a brief on the programme from the relevant UN official in charge of the program. In this briefing we were also joined by the in-coming Canadian contingent and we sat across from one another in the briefing room. After the briefing concluded, questions were invited from both contingents. The first couple of questions were answered ok, and then questions began to become a little more 'in-depth' concerning various aspects of how the training program was conducted. I recall watching the continual amazed looks as both the Australians and Canadians exchanged eye-rolling glances as questions were answered by the official with the same words: 'I don't have an answer for that question as I have no information regarding it' he would answer with a tight smile-and this was the man in charge.

The next problem the team was confronted with was where was the training package, you know, the base material where lessons would be devised and guidelines on time, content, duration of lessons, where conducted, whether practical or theoretical lessons, etc? Answer: there is no program you will need to make one up. Fortunately for us, the New Zealand government had also contributed a contingent to UNMCTT and they were a few weeks ahead of us into their deployment. In the true spirit of the ANZAC tradition we soon sat down with the Kiwis and devised a program that was very similar. The idea was to be doing the same training and not causing undue confusion with the Afghan refugees.'

Bob Kudyba, 1st Contingent

1st Contingent Members at Peshawar in 1989.
Back Row from Left: Imants 'Monty' Avotins, Anthony Smith, Craig Coleman, George
'Jock' Turner, Robert 'Bob' Kudyba.
Front Row from Left: Alan Palmer, Carl Chirgwin, Graham Costello, Paul Petersen.

'The security situation with Pakistan at the time was fairly stable; that is in major cities such as the nation's capital, Islamabad. Out in the North-West Frontier Province, where Peshawar was situated, this was not the case. Sentiments against 'Foreigners' varied from open signs of friendliness to open hostility. All the houses in Peshawar had high walls around them and the gates were manned by Chowkidars (Guards). The more wealthy and influential members of the Peshawar community had armed Chowkidars who touted the ubiquitous AK-47 rifle and a couple of spare magazines on a leather belt.'

Bob Kudyba, 1st Contingent

'Our neighbour in Peshawar was one of these wealthy locals and employed a couple of armed Chowkidars around his place during all hours. One day, one of these men had a conversation with the fellows looking after our residence and informed them that this gentleman was offended by people on the roof of our house without a shirt on. The water tanks on every house were situated on the roof areas of dwellings, water 'came on' during set hours of the day and night and one of the a chores was to set the pump going to fill these water tanks to ensure the house had adequate water for washing, showering cooking and toilets. It transpired our fellows used to climb right up there on the tanks to check water levels now and again, forgetting they were shirtless. It also transpired this gentleman's daughters' bedrooms looked out over our house.

The message passed was that if he continued to see shirtless men, he would start to shoot at them. Getting around in Peshawar in shorts was not on for men, nor was bare arms and singlets etc. A couple of members of the Aussie contingent went out for a jog, one had rocks thrown at him and then someone fired a shot at the other member for jogging in shorts. I think this also happened with a member of the French contingent who had a couple of shots put into the ground in front of him, when out jogging.

View from Australia House in Hayatabad in Peshawar in 1992.

Just to reinforce the message, we received some 'in-coming' a day
or so afterwards; Jock Turner and myself were standing in a room
near a window discussing a particular issue concerning training
when a few shots rang out, the 'crack-thump' of single 7.62mm
AK rounds whizzed past the corner of the building only about a
metre away. We both beat a hasty retreat out of the room.'

Bob Kudyba, 1st Contingent

'Being a good soldier, I went out for a jog. As you do, I wore
a t-shirt, shorts and runners. Trotting past an old bloke he
started throwing rocks at me. Guess he didn't like the colour
of my runners.'

Carl Chirgwin, 1st Contingent

'Travel to Peshawar was by commercial air with a two-night
stopover in Singapore. Sydney-based families waved goodbye
to us at SME (my mother and brother drove up from Canberra
for this) as we boarded a bus to Sydney airport. This was my
first real international travel and the sense of anticipation was
high. Airport processes were all a new experience, ably guided
by Ian Mahoney, who was a veteran of several overseas training
missions for SME.

Singapore was an eye-opener, but provided an unexpected opportunity to catch up with a Duntroon classmate, Goh Kee Nguan, who was a generous host to Bruce Murray and me as he showed us around his town. On the next leg to Karachi we started to get a feel for the world we were entering, starting with the spectacle of people being held on the airport tarmac under armed guard, followed for a few of us by a five-hour taxi tour of the city to fill in the wait for our connecting flight to Islamabad. With the benefit of hindsight that illustrated how little we understood the security situation in Pakistan at the time, but was a great experience that added to a building sense of adventure.

Arriving at Islamabad Airport late at night, we found ourselves confronted with another long wait for our onward (and final) regional flight to Peshawar. At that late hour, further taxi tourism wasn't an option so we settled down on the rather hard marble floor of the airport dining room to wait out the dawn. The final short flight in a Fokker Friendship was short, but provided a first glimpse of the terrain of rural Pakistan.

The delays we'd experienced en route had thrown out the 1ˢᵗ Contingent's reception plans and we emerged from Peshawar Airport to find no-one waiting for us. Once alerted to our arrival via a phone call, the 1ˢᵗ Contingent swung into action and quickly arrived en masse, complete with a highly modified Toyota Coaster 'flying coach' to convey us to our accommodation. This involved riding on the roof of the otherwise empty bus, allegedly to provide an air buffer against mine blasts, but really to give us a novel perch from which to experience our first sights of Peshawar. This was accompanied by explanations of our route and the sights along the way. By the time we arrived at the house we were to occupy in Hayatabad, via the Sadar market district and Jamrud Road, the sensory overload was pretty much complete, as intended.'

Andrew Smith, 2nd Contingent

2nd Contingent Members at Peshawar in 1990.
Back Row from Left: Graeme Toll, Ian Mahoney, Chris Reeves, Allan Mansell.
Front Row from left: Les Shelley, Bruce Murray, Bill Van Ree, Andrew Smith, Phil Palazzi.
(AWM P02121.001)

'One day half the students just didn't turn up for class. When we asked what was going on, we were told that there was going to be a big battle in Jalalabad and they had gone to fight. Sure enough, the battle happened and 'most' of the students came back to class afterwards.'

Bill Van Ree, 2nd Contingent

'The travel from Australia to Peshawar was a memorable and sometimes scary adventure in itself. It started well with a Qantas flight from Sydney to Singapore and a stopover there for a day or so and then a Singapore Airlines 737 took us from Singapore to Karachi. The plane was clearly the plane that was specifically assigned to that route by the airline back at the time. It was like a flying bus and once we were in the air for 30 minutes, the floors of the toilets were covered in urine and sodden toilet paper and the unique aroma of the inflight meal that passenger next to me was happily consuming had me standing in the isle with my head and arms against the overhead lockers above WO1 Phil Palazzi and dry reaching. I had a strong stomach and that was the only time that ever occurred, so well-done Singapore Airlines.

We arrived in Karachi at around midnight local time and after wandering around inside the airport for that appeared to have been modelled on a World War II Russian air-raid shelter we reluctantly lay on the floor as seating was almost non-existent and tried to grab some shut-eye. The airport security all armed with sub-machine guns would only open the main doors to the airport one hour before flights back at the time because of the high terrorist threat. So, we were stuck inside for a few hours and the had to hobble out with our luggage and wait away from the main building until re-entering one hour prior to our connecting flight to Islamabad.

That was another experience that had remained etched in my memory since. This leg saw us aboard a Pakistan International Airlines 747 that must have been the first prototype of the model. Local people boarded with what appeared to be pallets of cargo that were deposited in the aisles, not carry-on baggage stowed in overhead lockers. Almost none of the passengers wore seatbelts for the entire flight. When we touched down in Islamabad, we seemed to still be travelling at air speed. That was until we got a few metres from the end of the runway and the pilots snapped on every brake that the old bucket of bolts had and that in turn created and inertia effect that put everything inside the fuselage in forward motion. There were Pakistanis, large bundles of their carry on cargo and other miscellaneous items flying through the cabin, tumbling down the aisles, rolling off seats, the overhead lockers flew open and spilled their contents and the overhead cabinet that would normally contain the projector for the inflight entertainment but fortunately the projector had been removed in this case, swung down onto the seats below it. The passengers simply took to their feet once we had taxied to a stop, dusted themselves off and prepared to disembark.'

Allan Mansell, 2nd Contingent

'Our inductions in Peshawar commenced as soon as we hit the deck. The 1ˢᵗ Contingent had prepared an extensive and very thorough handover for us. Staff Sergeant Ian Mahoney and I were EOD Technicians and were familiar with most of the types of ordnance that we would be training the students on. However when it came to landmines back then, the pair of us and the rest of the team were up to speed on our own service mines, but none of us were familiar with the wide range of Soviet, Italian, Yugoslav, Czech, Pakistani and Iranian landmines that we would be focussing on for the training. So, over the ensuing days we received a crash course from the outgoing team members on every type of landmine, fuse and item of ordnance that we would be instructing on. It was a matter of familiarizing and adapting ourselves to their manufacture, categories, actuating principles, effects and methods for laying, neutralizing and disarming them. We also received a crash language course on Pashtun and Farsi greetings, terms to use during instruction, counting and the basic stop, go, move closer, move away, etc. The handover was very professional and that made the transition between the two contingents on the job as smooth as was possible given the short period that was available for the handover.'

Allan Mansell, 2ⁿᵈ Contingent

'My wife and I arrived in Karachi airport where the crowd of arriving passengers all mobbed the immigration desks waving their passports in the air above them. No cues or waiting behind the yellow line here. Well I thought, when in Rome..., so we plunged into the throng and forced our way towards an immigration officer. Our white faces—and my wife's pregnant belly—must have caught his eye as we managed to push our way towards him reasonably quickly. He was sweating profusely and stamped our passports officiously. We then went to collect our five large bags of luggage there was another scrum and wrestle to get it all through the crowds. Then despite almost everyone walking through customs we were asked to have our bags inspected—another 20 minutes as everything was unpacked and re-packed. By the time we got to the exit I was absolutely buggered.

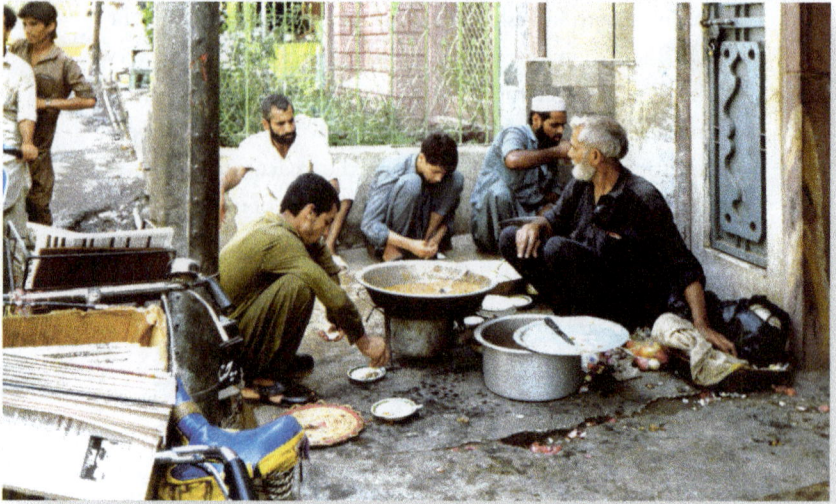

Sidewalk cafe in Peshawar in 1992.

'We had a pleasant night in a hotel before having to catch a flight to Islamabad. Checking in was a little easier but the 600m walk across the tarmac to the airport was tough as the temperature climbed. As we climbed the steps I noticed, with concern, two mechanics at work deep inside one of the engines. Thankfully, they soon finished whatever job they were performing but seemed to be having enormous difficulty in closing the cowling. One of the even gave it a kick but it just wouldn't shut. The other then looked back inside and pulled out a wrench that had been left in the way. The two of them smiled at each other, closed the cowling and walked away—only a minute before the pilot started the engines.'

Graeme Membrey, Technical Advisor

'We took over the house that had been used by the Canadian contingent. It came with seven household staff—some of who slept in a set of small rooms at the rear of the property. It was not easy to get used to having somebody always being there ready to offer assistance or serve you. The Canadians also left behind dozens of boxes of Cornflakes, hundreds of jars of peanut butter and thousands of packets of Ovaltine.'

Graeme Membrey, Technical Advisor

'Almost all of us got a dose of the shits as soon as we arrived. Not only was the food and water very different but the hygiene standards were pretty poor. Imodium became our close friend. As soon as you got a gurgle in your tummy you would whack down a few tablets and hope that you could head things off at the pass as it were. Sadly, the tablets quite often didn't work and you would end up spending a few days on the toilet with the 'squirts' and every so often you would get a dose that caused you to vomit at the same time.'

Anonymous, 4th Contingent

'It took a while to come to terms with a different culture and way of doing business. Not only the Pakistani way but the UN way as well. The Pakistani civilians in the program still displayed a lot of the vestiges from the British Empire time. Lots of bureaucratic process and paperwork. Everything just seemed to take time and there was no real sense of urgency. We all said 'Inshallah' [if Allah is willing] in the vague hope that something would actually happen. I guess they were there for the long haul and we were just passing through.'

Anonymous, 4th Contingent

Catching a bus in Peshawar in 1992.

Selling dyed baby chickens at a market in Peshawar in 1992.

'As our Pakistan Airlines jet approached the Peshawar airstrip it flared for landing. The sun was setting and the rugged mountains through which the Khyber Pass makes its winding way, took on a warm red hue. Those close to the windows glimpsed pieces of the frontier town of Peshawar which was to become our home. Without warning we were driven into our seats as the pilot pushed the throttles to the stops and the cabin filled with the whine of jet turbines scrambling for altitude. We repeated this process three times and by now the sun had set. Finally, the pilot put mouth to microphone and advised we wouldn't be landing in Peshawar as there was too much gun fire over the airport. The gunfire was not aimed at the plane. It was customary for men to fire their weapons into the air at certain celebrations and this was probably the case. Stories abounded of revellers suffering an untimely demise when the reveller behind pulled the trigger a fraction too early. This was our welcome to Pakistan.

The plane was rerouted to Karachi for refuelling. On arrival the cabin doors were thrown open and refuelling commenced, no problems there except we were still on the plane and all the smokers congregated at the doors flicking their cigarette butts on the tarmac in vicinity of the refuelling truck. Call me overly sensitive, but as I was sitting in my seat watching this happen when the lyrics of a popular Don Maclean song ran through my head, 'this'll be the day that I die'. We took off again and landed in Islamabad where we were met by Warrant Officer Danny Hawkins from the 4th Contingent. His presence seemed more like a rescue than a welcome.'

Brian O'Connell, 5th Contingent

'We were just about ready to go by this stage so we painted our details on our security trunks and took them to the air freight terminal at Sydney Airport to be sent as unaccompanied baggage, said our farewell and headed off to Peshawar.

The first leg was Qantas to Singapore in Business Class. We were in the little upstairs 'bubble' of the jumbo jet which only had about 20 passengers so we got chatting with the steward

Removing an illegally parked car with a forklift at Peshawar Airport in 1992.

5th Contingent Members at Peshawar in 1991.
From Left: Brenton White, Michael McQuinn, Michael Kavanagh, Brian O'Connell, David Edwards, Dean Brown.

as he was bringing out the beers. When the steward found out where we were going and what we were doing, he just threw us the keys to the beer fridge and said to help ourselves. Either that, or he got tired of bringing out beers at the rapid rate.

We stayed at a very nice hotel in Singapore which was opposite the famous Raffles Hotel, which was good as I wanted to have one of their famous Singapore Sling cocktails. This was only my second time out of Australia (the first being to Vanuatu to build a patrol boat base with 17th Construction Squadron a few years earlier) and growing up in country NSW I had read about the Australians in Singapore during World War II and how Raffles was a popular night spot, obviously before the Japanese arrived. Unfortunately, Raffles was closed for renovations so I missed out on my cocktail.

As Sydney time is a few hours ahead of Singapore, I was up early the next morning and went for a walk around the city. It was really quite moving thinking about the young Australian soldiers who had been sent there in world War II only to end up in captivity.

We were travelling again that day from Singapore to Karachi via Bangkok. We flew Thai Airlines on both legs and the flight to Bangkok was uneventful but boarding the flight to Karachi was a real eye opener. Some of what the passengers were trying to carry on had to be seen to be believed. It looked like one guy was trying to carry on a clothes dryer while another looked like he was trying to carry on a replacement front end for his car. Needless to say, the Hosties weren't going to allow this so boarding took quite some time to sort out.

It was dark when we arrived into Karachi so we didn't get to see much from the aircraft. Going through Pakistani Customs

The infamous unsafe bridge on the Grand Trunk Road in 1992.

The Headquarters of the Frontier Corps at Peshawar in 1992. Note the two tuk tuks.

and Immigration was almost as interesting as boarding in Bangkok. I remember there were lots of people and quite a lot of noise with about 20 or more lanes for Pakistanis returning home with only two or so lanes for foreigners. The official sat up quite high, like a tennis umpire, so we had to hand up our passports for scrutiny. After a while looking at the passport and looking at us, the official picked up this iron stamp with a huge timber handle that looked like it had been in use since Queen Victoria's time and smashed it down on our passports and we were officially in Pakistan.'

Michael Kavanagh, 5th Contingent

'Danny Hawkins from the 4th Contingent was there to meet us when we came through customs which was good as there were even more people outside the terminal than inside and they all wanted to help us carry our bags, for a fee of course. Danny did a good job keeping them away from our luggage while also guiding us to the minibus he had arranged to take us to the hotel.

I don't remember much about the hotel but we were off to Peshawar the next day and I remember that it was a Friday which I will explain later. On the bus back to the airport, we got to see some of Karachi as it was during the day and there were two things I remember about the traffic. The first one was the number of small motorcycles and how many people they fitted on them. I can remember seeing one with dad driving (of course), mum sitting side saddle on the back carrying a baby, a young boy sitting on the petrol tank in front of dad and another child sitting between mum and dad making five people on a Honda 125 road bike. The other thing I remember is the complete disregard for the number of lanes on the road. At traffic lights, cars fill the available lanes while others push in beside them so there are at least twice as many cars lined up as there are lanes. When the lights go green it is like the start of a race as they all jostle for position heading down the road.

The flight to Peshawar was with Pakistan International Airlines and I remember the aircraft being quite dirty with finger prints all over the windows and the little napkin things on the headrests being all brown and looking like they hadn't been changed in months. I remember thinking that I hope they pay more attention to maintenance than they do to cleaning.

We arrived over Peshawar just on sunset with the sun setting over the Hindu Kush Mountains providing an impressive backdrop. I was looking out the window at our new home for the next few months and we were getting quite low when the pilots suddenly engaged full thrust and started climbing again. They were also pushing the Hostess call button quite a lot which I didn't think was a good thing as the Hostess was trying to claw her way forward as the aircraft continued to climb. I thought we were going around to have another go at landing in Peshawar but after a while the announcement came on that 'Inshalla, we will be landing in Islamabad' and, sure enough, we did.

We sat on the aircraft for quite a while not knowing what was happening and then we got told that everyone had to get off the aircraft so we sat in the terminal for a while, still not knowing what was happening. We then saw the pilots walking past so we asked them what was going on. They advised that when they were trying to land in Peshawar, someone was firing their automatic weapon into the air right in front of the aircraft so they had aborted the landing. They said this happens all the time on Friday's in Peshawar when people celebrate things like weddings by firing randomly into the air. They also told us they weren't going back to Peshawar that night (not sure why) so we would have to stay in the terminal until tomorrow. We didn't think this was a very good idea so grabbed our luggage (which was quite a bit) and headed out to find some taxis.

The airport is actually in Rawalpindi and the taxis looked like they were Morris Minors from the 1950s, and probably were, and I believe they were green with a yellow roof. I remember three things about the taxis. The floor where I was sitting was rusted out. Not a little bit, but a lot. The second thing was that there was something wrong with the exhaust and quite a lot of it was coming into the cabin which smelt like a two stroke it was burning that much oil. The third thing is that they aren't very big and we had a lot of luggage, even with two Aussies per taxi, so our luggage was hanging out all over the place.

So off we head to the hotel in a fleet of 1950s Morris Minor taxis and someone decided it would be fun to have a race. I'm not sure who won but we all managed to arrive safely at the Pearl Continental Hotel in Islamabad which had the accommodation contract for the Australian High Commission. It was also blown up in the more recent troubles in Pakistan and Afghanistan, but in 1990 it was very nice.

The next day, a mini bus came down from Peshawar to collect us and off we went to experience our first trip on the Grand Trunk Road and commence our handover with the 4th Contingent. The Aussie house was in Hayatabad which was the modern part of town.

The 4th Contingent was is a two storey, six-bedroom, six-bathroom house that was rented by the UN so we paired up with our opposite numbers and moved in. The 4th Contingent was keen to see us and threw a belated party for our belated arrival. We were advised that they were waiting for us at the airport with a special 'Demining' Welcome to Peshawar experience when they saw the aircraft climb away and not return. The 4th Contingent was dressed in their Pathan outfits and had rented tuk tuks for us to travel through Peshawar which would have been a memorable way to arrive but was not be.

I was taking over from Ken Norman who had been the Operations Officer on Deming Headquarters. When the program was first established there were quite a few contributing nations but by late 1990, most of these had departed leaving only the Kiwis in Quetta and the Aussies and Americans in Peshawar. The American contribution was based on a 12-man Green Beret Team from the 5th Special Forces Group based at Fort Bragg, North Carolina with Captain Chris Papas commanding. Chris had been the Administration Officer on Demining Headquarters and, now that Ken was departing, became the Operations Officer through some deal Ken and Chris had resolved as Chris had more experience in Pakistan, with me as the Administration Officer. So, Ken handed over to Chris and Chris handed over to me.

I could see why Chris didn't want to be the Admin Officer any more as it wasn't a particularly interesting role and amounted to ordering materials for training, ordering equipment to be issued to the graduates and authorizing invoices to be paid. In addition, there was a Pakistani employee (Farhad) who did most of the paper work. Fortunately, I had a number of other roles being: Technical Advisor to the Organisation for Mine Awareness (OMA); Technical Advisor to the Mine Clearance Planning Agency (MCPA); and Contingent Second in Command.

After Chris provided me with the finer points on UN paper trails and all the signatures required for payments to occur, Ken rescued me and took me to meet the heads of OMA and MCPA. OMA provided training in Mine Awareness and was led by Faisal Kareem who had studied in America and fought against the Russians. He was in his 40s and was quite tall being well over 6 foot. MCPA undertake surveys of the minefields and areas of unexploded ordnance was led by Sayad Akar who had also studied in America but I'm not sure what his role was in defeating the Russians. Sayad was younger than Faisal and was in his late 20s or early 30s.

Ken then introduced me to one of the most important people I would need in my time in Pakistan and that was our driver, Numnar Khan. I'm not sure when Numnar joined the Demining Program but he told me in his limited English that he had driven for Bruce Murray as well as Ken and I found him to be reliable and helpful but, most importantly, he knew where everything was and how to navigate his way there. Prior to arriving in Pakistan, I had visions of travelling around in a white UN Toyota Landcruiser. The white Toyota parts were correct but instead of a Landcruiser we had a Corolla.

So Ken and Numnar took me back down the Grand Trunk Road on a two day tour of Islamabad to introduce me to the Australian High Commission staff, get our team registered at the UN store in Rawalpindi and introduce me to some people at the British High Commission who were able to help us out with Australian beer. Being a Muslim country, alcohol was a highly controlled substance. Being part of the UN, we could purchase alcohol but only through the UN store. To do this, we had to be registered and pay a $US200 bond for each membership. This allowed us to purchase up to a monthly quota that the UN had negotiated with the Pakistani government and covered spirits, wine and beer, with one membership allowed to purchase four cartons of beer per month. $US200 was quite a lot of money back in 1990 so Ken advised that his team had taken out three memberships allowing 12 cartons of beer per month. This sounds like a lot for 6 people, but Ken pointed out that there were a lot of westerners in Peshawar with the various programs who couldn't purchase alcohol and were looking to us to assist. Interestingly, bacon was also a prohibited substance in Pakistan so we could purchase cans of bacon from the UN store as well.

After registering at the UN store and buying our initial supply of European beer, and sorting out the extensive amount of paperwork associated with the transaction of controlled substances, we loaded up the Corolla and headed to the British Club for lunch. The British High Commission in Pakistan was, according to those who worked there, the second largest British mission after the USA. As such, their club was well stocked, well run and well patronised. The British also had a common sense approach to security so it was fairly easy for Aussie Deminers from Peshawar to get into the club. They also imported beer from Australia by the container every few weeks.

After meeting up with a few of the British High Commission staff, it was back to Peshawar with Ken pointing out where all the 'western' toilets were located between Peshawar and Islamabad. Middle Eastern countries have an open pit/ squatter type of toilets that are only marginally better than just going in the bush. In addition, diarrhea was an ever-present risk so it was important to know all the Tea Houses where they had good, clean, western style toilets in case you needed to use one in a hurry. The 4th Contingent also advised us to carry the lomitol tablets from the medical kits with us at all times along with a clean pair of underpants, just in case.

With hand overs completed, the 4th Contingent members departed and it was now all down to us. I remember thinking that Peshawar, the UN and the Mine Clearance Program was a bit like being in a Jason Bourne movie where I didn't know what was true, what wasn't and what I was being told to make me think a certain way.'

Michael Kavanagh, 5th Contingent

'Around half way into our tour one of our members had to return home for personal reasons. His replacement Ben White was due to arrive at the Peshawar airport shortly thereafter. Around our time the Gulf War was pretty well in full swing and the locals were edgy. Some riots and protests made for interesting days and the other supporting deminers had left under the cover of darkness once things started heating up. We decided to welcome our new recruit with a good old gee up to break the ice so to speak. No doubt he would have been brought up to speed with the current situation but I'm sure he wasn't ready for what we planned. Waiting for him at the arrival area Barry and myself quickly covered his head and with great haste shuffled him into the back of our van telling him the whole time to keep low and remain inconspicuous. The van driver hurtled us through back streets and bazaars in what would have been like something out of Fast and Furious along with screaming to get us home post haste. All the while our newly arrived was pushed into a crease of a seat and told

A typical decorated truck in Pakistan—known as a jingly truck because of the sound the decorative chains make.

to keep his head down. This went on for quite some time until finally we approached our home. We told Ben to remove his head gear and on two wheels, smoke and screeching came to a halt on our driveway to be greeted by the remaining members. Now not only was he born a White, that day he resembled it.'

Dean Brown, 5th Contingent

'After we started working in Afghanistan, I started the process of seeking some additional hazardous condition allowance for the team. The Defence Advisor in Islamabad—Brian Cloughly— supported the application by commenting 'Their domestic circumstances while living in Peshawar can be described, generously, as isolated, tedious and to a degree hazardous. Their conditions would excite the admiration of a native-born Spartan. It may be imagined that conditions within Afghanistan are even less comfortable and even more dangerous.'

Warren Young, 7th Contingent

The rare sight of a woman shopping at Peshawar in 1992.

'On the domestic flight from Karachi to Peshawar I was amused to hear the pilot say over the PA 'Inshallah (if Allah is willing) we will soon be taking off'. When we did take off the jet climbed at about 45 degrees—the pilot must have been a former fighter jock. And one of the passengers who had clearly never flown before panicked, jumped up out of his seat and tried to run to the front of the plane. He fell flat on his face and the air hostess just looked at him like she had seen this happen a hundred times. He wasn't able to get back into his seat until we levelled out. When our plane landed at Peshawar airport a ring of armed soldiers formed up around the aircraft. Jeez, I thought, this looks it might get interesting. But as it turns out it was completely routine. Getting off the domestic flight in Peshawar the push and shove to recover your bags and then out to find your ride was like being in a rugby maul.'

Marcus Fielding, 8th Contingent

'The thing that suddenly struck me after a few minutes driving through Peshawar was the absence of women. And those that you did see were dressed in burqas—usually blue or black in colour. I learned that women are generally kept in the home and are only allowed out in certain circumstances. They weren't allowed to be out alone and weren't allowed to drive. It turns out that men do any shopping that might be needed for the household. Later I learned that Peshawar was relatively liberal; women in the tribal areas are

At a market in Peshawar in 1992.

8th Contingent Members at Peshawar in 1992.
Back Row from Left: Barry Veltmeyer, Dean Beaumont, Clyde Jochheim,
Darrell Crichton, Danny Shaw.
Front Row from left: Mark O'Shannessy, Marcus Fielding, Rex Wright, George O'Callaghan.

not permitted to go outside the household compound. Girls were married off pretty much after they started menstruating.'

Marcus Fielding, 8th Contingent

'One of the first things to arrange on arrival was to get a tan uniform (that looked very much like the Pakistani Army uniform) tailored. The shirt and trousers were ready overnight but they weren't very well tailored. With that uniform we wore a brassard with an Australian Kangaroo badge and a UN badge on them. We also went down to the local bazaar to buy a couple of sets of shalwar kameez (loose trousers and long shirt), a vest jacket (with lots of pockets) and a 'Chitrali pakool' hat. We were advised to get dark colours like grey, blue and brown. Not white! The tan uniform was for wearing in Pakistan and the mufti was for when we went into Afghanistan.'

Marcus Fielding, 8th Contingent

Well decorated and well loaded trucks in Pakistan—known as jingly trucks because of the sound the decorative chains make.

'Shooting weapons up into the air either to test fire them or to express your joy—particularly at a wedding—was a local custom you don't find in too many parts of the world—but in Peshawar it happens all the time. When the Pakistani cricket team won the World Cup it sounded like World War 3 had started! But the curious thing was that no one seemed too concerned that anything that goes up must come down. A number of people were killed or injured when rounds returned to earth. You would sometimes come across fired rounds on the ground and there was even a report of one hitting one of our vehicles. We had also heard that the 5th Contingent's plane coming into Peshawar had to be diverted because of the amount of firing going on—it must have been a big wedding!'

Marcus Fielding, 8th Contingent

'Upon arrival in Peshawar we were ushered into tuk tuks and thrown into the mayhem of local traffic, fumes and smells, sounds and colours. Adding to our sense of disorientation, a welcome party followed at one of the Aussie Houses. Our local staff provided the music (drums and wooden flute) and proceeded

Street scene at Peshawar in 1992.

Afghan refugee dwellings in Peshawar in 1992.

to encourage us into the dance circle where we placed arms on shoulders, and commenced to dance and clap the afternoon away. In typical fashion though, our memories faded of our own experiences of discomfort when the forward elements of the 10th Contingent arrived. While the cultural aspects of the welcome party, music, and dance circle remained unchanged, we decided not to expose them to the tuk tuk ride from the airport; instead we provided them with donkeys. I think that's called jack!'

Craig Egan, 9th Contingent

"Recoilless'—an all too common term used between Contingent members when politely explaining ones unfortunate and repeated ailment causing the violent venting of expanding gas while remaining motionless.'

Craig Egan, 9th Contingent

'Soon after our arrival we visited the ICRC hospital in Peshawar to improve our perspectives on the tragic impact land mines and the remnants of war were having on the population. Serving to orientate our understanding of the broader UN and NGO efforts, and familiarise ourselves with the quality and accessibility of treatment available to refugees, this first visit remains vivid in my mind. A limited stockpile of prosthetic limbs was available, with a craftsman refining the shape to their identified recipient's needs.

While being hosted by the ICRC volunteers, we met an elderly Afghan man standing on his newly fitted prosthetic leg; he was

Afghan children commuting on a camel in Quetta in 1992.

about to be discharged. What was remarkable is the fact that not less than 10 days prior this steely eyed proud old man had lost his lower leg to a land mine. From a remote, rugged and mountainous Afghanistan border area, with assistance only from family and friends, this elderly gentleman had endured four days of travel, a mixture of being carried, stretchered, or draped over a camel or donkey. No pain relief, not a whimper or complaint, just pure determination and a deep trust in those around him. This, gentlemen, remains the most remarkably resilient, courageous, humble and stoic person I have ever met, and forged my perception of the Afghan character, people and Nation.'

Craig Egan, 9th Contingent

Pashtuns

The Pashtuns, historically known as ethnic Afghans or Pathans are an Indo-Iranian ethnic group native to South-Central Asia, who share a common history and culture. A substantial majority of ethnic Pashtuns shares Pashto, an Eastern Iranian language in the Indo-European language family as the native language.

Globally, the Pashtuns are estimated to number around 50 million, but an accurate count remains elusive due to the lack of an official census in Afghanistan since 1979. The majority of the Pashtuns live in the region regarded as 'Pashtunistan', which has been split between two countries since the Durand Line border was formed after the Second Anglo-Afghan War.

There are also significant Pashtun diaspora communities in Sindh and Punjab in Pakistan, especially in the cities of Karachi and Lahore, and in the Rohilkhand region of Uttar Pradesh, India. A recent Pashtun diaspora has also developed in the Arab states of the Persian Gulf, primarily in the United Arab Emirates.

The Pashtuns are a significant minority group in Pakistan, where they constitute the second-largest ethnic group or about 15% of the population. As the largest ethnic group in Afghanistan (anywhere between 42 and 60 percent of the population), Pashtuns have been the dominant ethno-linguistic group for over 300 years.

The Pashtuns are the world's largest segmentary lineage ethnic group. Estimates of the number of Pashtun tribes and clans range from about 350 to over 400. There have been many notable Pashtun people throughout history: Ahmad Shah Durrani is regarded as the founder of the modern state of Afghanistan, while Bacha Khan was a Pashtun independence activist against the rule of the British Raj.

Pashtun culture is mostly based on Pashtunwali and the usage of the Pashto language. Pre-Islamic traditions, dating back to Alexander's defeat of the Persian Empire in 330 BC, possibly survived in the form of traditional dances, while literary styles and music reflect influence from the Persian tradition and regional musical instruments fused with localised variants and interpretation. Pashtun culture is a unique blend of native customs with some influences from South and Western Asia. Like other Muslims, Pashtuns celebrate Ramadan and Eid al-Fitr. Some also celebrate Nouruz, which is the Persian new year dating to pre-Islamic period.

Pashtunwali (or Pakhtunwali) refers to an ancient self-governing tribal system that regulates nearly all aspects of Pashtun life ranging from community to personal level. One of the better-known tenets is Melmastia, hospitality and asylum to all guests seeking help. Perceived injustice calls for Badal, swift revenge. Many aspects promote peaceful co-existence, such as Nanawati, the humble admission of guilt for a wrong committed, which should result in automatic forgiveness from the wronged party. These and other basic precepts of Pashtunwali continue to be followed by many Pashtuns, especially in rural areas.

Another prominent Pashtun institution is the loya jirga or 'grand council' of elected elders. Most decisions in tribal life are made by members of the jirga, which has been the main institution of authority that the largely egalitarian Pashtuns willingly acknowledge as a viable governing body.

Afghan Refugees

Afghan refugees are nationals of Afghanistan who left their country as a result of major wars or persecution. The 1979 Soviet invasion of Afghanistan marks the first wave of internal displacement and refugee flow from Afghanistan to neighbouring Pakistan and Iran that began providing shelter to Afghan refugees. When the Soviet war ended in 1989, these refugees started to return to their homeland. In April 1992, a major civil war began after the mujahideen took over control of Kabul and the other major cities. Afghans again fled to neighbouring countries.

A total of 6.3 million Afghan refugees were hosted in Pakistan and Iran by 1990. As of 2013, Afghanistan was the largest refugee-producing country in the world, a title held for 32 years. Afghans are currently the second largest refugee group after Syrian refugees. The majority of Afghan refugees (95%) are located in Iran and Pakistan. Some countries that were part of the International Security Assistance Force (ISAF) took in small number of Afghans that worked with their respective forces. Ethnic minorities, like Afghan Sikhs and Hindus, often fled to India.

There are over one million internally displaced people in Afghanistan. The majority of the IDPs in Afghanistan are as a direct and indirect result of conflict and violence, although there are also reasons of natural disasters. The Soviet invasion caused approximately 2 million Afghans to be internally displaced, mostly from rural areas into urban areas. The Afghan Civil War (1992–96) caused a new wave of internal displacement, with many Afghans moving to northern cities in order to get away from the Taliban ruled areas. Afghanistan continues to suffer from insecurity and conflict, which has led to an increase in internal displacement.

After the removal of the Taliban regime in late 2001, over 5 million Afghans were repatriated through the UNHCR from Pakistan and Iran to Afghanistan. Hundreds of thousands of Afghans began returning to Afghanistan in recent years. According to the United Nations, by the end of 2016 about 600,000 documented and undocumented Afghans were repatriated from Pakistan. According to the IOM, the return of undocumented Afghan refugees from Pakistan in 2016 were more than twice the number of 2015, increased by 108 per cent from 2015. The remaining registered Afghan refugees in Pakistan numbers around 1.3 million. In the same year, UNHCR reported that 951,142 Afghans were living in Iran. Most of them were born and raised in Pakistan and Iran in the last three and a half decades but are still considered citizens of Afghanistan.

Mujahideen

Mujahideen is the plural form of mujahid, the term for one engaged in Jihad (literally, 'striving' or 'struggling', especially with a praiseworthy aim).

Its widespread use in English began with reference to the guerrilla-type military groups led by the Islamist Afghan fighters in the Soviet–Afghan War, and now extends to other jihadist groups in various countries.

Arguably the best-known mujahideen outside the Islamic world, various loosely aligned Afghan opposition groups initially rebelled against the government of the pro-Soviet Democratic Republic of Afghanistan (DRA) during the late 1970s. At the DRA's request, the Soviet Union brought forces into the country to aid the government from 1979.

The mujahideen fought against Soviet and DRA troops during the Soviet—Afghan War (1979–1989). Afghanistan's resistance movement originated in chaos and, at first, regional warlords waged virtually all of its fighting locally. As warfare became more sophisticated, outside support and regional coordination grew.

The basic units of mujahideen organization and action continued to reflect the highly decentralized nature of Afghan society and strong loci of competing mujahideen and tribal groups, particularly in isolated areas among the mountains. Eventually, the seven main mujahideen parties allied as the political bloc called Islamic Unity of Afghanistan Mujahideen.

Many Muslims from other countries assisted the various mujahideen groups in Afghanistan. Some groups of these veterans became significant players in later conflicts in and around the Muslim world. Osama bin Laden, originally from a wealthy family in Saudi Arabia, was a prominent organizer and financier of an all-Arab Islamist group of foreign volunteers; his Maktab al-Khadamat funnelled money, arms, and Muslim fighters from around the Muslim world into Afghanistan, with the assistance and support of the Saudi and Pakistani governments. These foreign fighters became known as 'Afghan Arabs' and their efforts were coordinated by Abdullah Yusuf Azzam.

Although the mujahideen were aided by the Pakistani, US, and Saudi governments, the mujahideen's primary source of funding was private donors and religious charities throughout the Muslim world—particularly in the Persian Gulf. Journalist Jason Burke asserts that 'as little as 25 per cent of the money for the Afghan jihad was actually supplied directly by states.'

Mujahideen forces caused serious casualties to the Soviet forces, and made the war very costly for the Soviet Union. In 1989 the Soviet Union withdrew its forces from Afghanistan. Many districts and cities then fell to the mujahideen; in 1992 the DRA's last president, Mohammad Najibullah, was overthrown.

However, the mujahideen did not establish a united government, and many of the larger mujahideen groups began to fight each other over power in Kabul. After several years of devastating fighting, a village mullah named Mohammed Omar organized a new armed movement with the backing of Pakistan.

This movement became known as the Taliban ('students' in Pashto), referring to how most Taliban had grown up in refugee camps in Pakistan during the 1980s and were taught in the Saudi-backed Wahhabi madrassas, religious schools known for teaching a fundamentalist interpretation of Islam. Veteran mujahideen confronted this radical splinter group in 1996.

Risalpur

Risalpur is a city in the Nowshera District of Khyber-Pakhtunkhwa, Pakistan, on the Nowshera-Mardan Road. It is 45 km from Peshawar.

Risalpur is located in a basin some 1014 feet above sea level, it is bounded on the south and west by the Kabul and Kalpani rivers, respectively. The Risalpur Military Cantonment itself lies on high ground, some 30 feet above the surrounding area, north of Nowshera and south of Mardan.

The Risalpur Military Cantonment hosts several Pakistan Air Force and Pakistan Army units and schools including the Pakistan Air Force Academy and the Military College of Engineering. Risalpur is known as 'Home of Eagles' and 'Home of Sappers'.

Chapter Five
Working in Pakistan

The first five contingents worked at Risalpur—some 45 km from Peshawar. They would usually have a very early start to drive to Risalpur and start training early in the morning. Because it got so hot the training usually finished up around lunch-time. The contingent members would then return to Peshawar. The working week was Sunday through to Thursday as Friday was the Muslin holy day.

Courses were run in three languages—Pashto, Urdu and Dari—to accommodate the student body. Even then, some students weren't in a position to understand any of these languages. Often the students were illiterate so they were not able to read or do any sums. This presented a problem in determining the length of time fuze to be attached to an explosive demolition charge.

'The demining program had been initiated by the ex-Afghan Army Colonel Kefayatullah Eblagh. To his credit he attracted considerable resources from the UN, Pakistani Military, and overseas countries. It was his role to manage the complex political situation. Amazingly given the past decades of upheaval, the Afghan Technical Consultants has grown in capabilities and still operating in Afghanistan. He personified the very best in his culture's hospitality and politeness. One could spend all day arguing with him as to who was going to pass through the door first. 'No, you, no you first'. Eventually, we would have to give up and go through the door first. We were incredibly naïve about cultural norms. Kefayatullah once visited our team house. It was very hot on the day and to my everlasting regret I greeted him in shorts and shirt, whereas we should have rushed off and dressed formally covering up.

Kefayatullah often introduced us to leading mujahidin figures. Many went on to be famous and/or notorious depending on your point of view. Towards the end of our engagement, Kefayatullah invited me for lunch at his house. I was ushered from the bright sun light into a modest mud brick building and into a very dark room. As my eyes accustomed to the darkness, I found there were about fifteen leading mujahidin leaders in the room. All were covered in ammunition belts and carrying AK-47 assault rifles. All were clearly Afghan except for a very tall man dressed in white who I was told was from Saudi Arabia. We sat on an ancient Afghan Filpai (elephant's foot motif) carpet of dark reds and blues that filled the room. A parade of Afghan dishes was brought into the room. There was much discussion of which my beginner Pashtu was not helpful. I consoled myself to enjoying the food and making an effort to at least try everything even if it didn't look as though it would have passed an Australian food test.

We were determined to rectify the lack of knowledge about mines and ordinance in Afghanistan. Apart from being essential information in the day to day running of demining, we wanted to raise the level of understanding in the wider Australian Army. With the shiny new (expensive) computer we set about writing detailed documents on the mines we found. We also stared to collect and record information from our students on how the Soviets and the Mujahidin had deployed mines in Afghanistan. This work resulted in a multi volume book that detailed every mine that had been deployed in Afghanistan, its technical specifications, deployment, and suggested demining. The books were supported by photos. Carl and Paul led the book production. Arnie Palmer led the initiative to film each of the contingent providing training briefs on each of the mines.

In our spare time, the contingent set about collecting examples of everything we could get our hands on. The mines would then be used as practice mines in the training. Visits to local arms bazars were the key sources we relied on. If the arms traders didn't have something, then we could put in an order. A couple of weeks later whatever we asked for would turn up. During our three months we collected samples of every mine deployed in Afghanistan. Most were Soviet but some were form other conflict zones. Many of the newer Soviet mines had never been seen before. Our team of qualified explosive ordinance disposal experts set about the dangerous task of disarming the mines and removing the explosives. Although risky I had total confidence in the team. All the same we took precautions just in case. The American team of special forces unfortunately were not as well qualified. Some of them were inspecting a mine in their team house when it exploded. Fortunately, they only had minor injuries but some of them had to be medevac'd home.

Of particular pleasure was the humour of the Afghans. We found that the Australians and Afghans had a similar sense of humour. Any opportunity for some humour or a joke was taken. The Afghan students had a fatalistic view of the future. Whilst serious students, they took every opportunity to enjoy the time they had. In their view, the future was in Allah's hands. It is hard to imagine that many of them have survived the decades of turmoil since the demining training.'

Graham Costello, 1st Contingent

'I was asked to go and have a chat with the 'engineer' from one of the mujahideen groups. They had discovered and finally managed to recover something they had never seen before. I didn't recognise it either but we took some photos and measurements. Later we were told that it was a VP-12/13—a nasty little device that controls the electric detonation of several anti-personnel mines based on the triangulation from two geophones. The reason that it had taken so long to recover one of these was that everyone who had tried was killed. The one that had been recovered had apparently malfunctioned or the battery had run flat.'

Carl Chirgwin, 1st Contingent

'One day I was asked to help identify the remnants of a SCUD missile that had apparently been recently fired over the border from Afghanistan to somewhere near Peshawar.'

Carl Chirgwin, 1st Contingent

'We never really worked out how it got there or why but one day the Kiwis noticed a disturbed area on the path between the firing point and the target area. It turned out that there was a freshly planted anti-personnel mine there...'

Carl Chirgwin, 1st Contingent

'In the early days were still casting around for ideas on how to achieve wide area mine clearance with high degrees of assurance. One of the British fellows in Quetta put forward a suggestion to use marksmen to fire at the mines. Of course, the mujahideen had been doing this throughout the war to get out of mined areas. But to achieve humanitarian standards i.e. 99% assurance it wasn't going to be good enough.'

Paul Petersen, 1st Contingent

'Nearly every male refugee claimed to be Mujahideen. It didn't matter if they were eight or eighty years old. Being a warrior was a badge of honour. I understand that Osama bin Laden lived in Peshawar in the late 1980s and established Al-Qaeda there. But he wasn't advertising for recruits in the local paper. Some of the Afghan instructors at our camp had no desire to go back to Afghanistan to clear landmines. They made comparatively good money as an instructor—about 4,000 rupees per month. That was a lot more than a civil engineer could earn in Afghanistan.'

Paul Petersen, 1st Contingent

'The senior military officer for the Demining Program in Peshawar was a US Army Lieutenant Colonel Larry. He was the most indecisive and incompetent officer I have worked with. He would sweat when faced with simple decisions, and was incapable of dealing with anything serious. I have always assumed he was sent to Peshawar by the US Army because it was the most remote posting they could find. The local UN headquarters was a model of grinding bureaucracy. The Chief

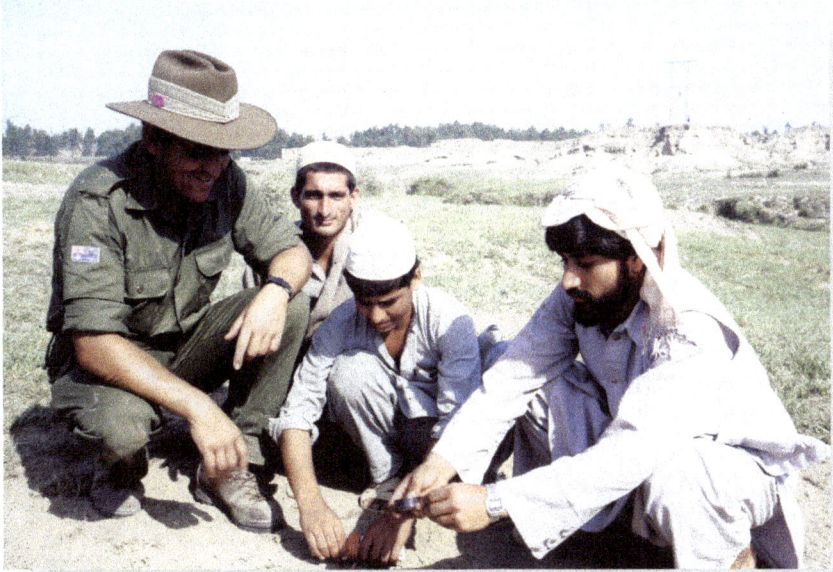

Imants 'Monty' Avotins (left) instructs Afghan students on the use of explosives at Risalpur in 1989.

of Staff would neither decide nor delegate. He only knew how to delay and dither. There was a lot of tension at the headquarters. They weren't used to Australians.'

Paul Petersen, 1st Contingent

'I think our Australian instructors were excellent. It wasn't just their knowledge; it was their ability to engage the trainees. Some of the other teams were patronising, but not ours. Some of the training teams from other countries were so bad that their interpreters did most of the training for them. The Americans were professional, but they were Special Forces, and I always thought there was a deeper agenda attached to their work. They could speak Pushto and Farsi, but that didn't mean they could relate to their trainees.'

Paul Petersen, 1st Contingent

'We had bugger all equipment and the mine detectors we had were pretty average. There was no shortage of gun runners trying to flog kit to the UN. On one occasion, we received a shipment of 'treasure finder' mine detectors from the

American company Radio Shack. They could detect coins etc., and were toy detectors marketed at children. Try telling that to UN Headquarters or Lieutenant Colonel Larry L. They thought we should be using them operationally. I was stunned. I can't imagine what Monty Avotins or Bob Kudyba would have said if they had known. I sent the treasure finders back to Islamabad with a rude note. Lieutenant Colonel Larry was furious with me.'

Paul Petersen, 1st Contingent

A group of Afghan students learning to use a metal detector at Risalpur in 1992.

'The first contingent was very much challenged with a steep learning curve as we arrived and were expected to get cracking into a project that nobody had really sat down and thought through...so we were literally making the rules of the training up as we went along.'

Bob Kudyba, 1st Contingent

'As there was little to no guidelines on what to train the classes of Afghans on, we sat down and looked at the basic military doctrine on landmine breaching teams; these drills date back to the Second World War and were based upon the idea of manually clearing the ground out from a known 'safe' area-that is an area where mines were known no to be. Military clearance was also aided (in the case of countries abiding by the Geneva protocols on mine warfare) by the fact mined areas were to be marked by signs and also wire barriers. However, this protocol was not adhered to by those nations who were not signatories to the Geneva protocol, nor was the protocol observed by various dissident, or guerrilla groups.

Military breaching drills were also inclusive of protection parties so that the men undertaking the actual clearance didn't have to also concentrate on carrying a rifle and associated kit as well as clearing the ground. So, we dropped all the protection party stuff and just focused on the people doing actual clearance. This would need people to prod the ground across a front of say, one metre width, and then investigate any finds they had. It would be slow tedious stuff and they would also need to clearly show what area had been subjected to clearance and what was not done. So, the military idea of using rope, or tapes either side of the clearer were used.

At this stage the only way the ground could be investigated was via prodding, as there were no metal detectors available. As the Afghans were not trained in ordnance, it was decided to avoid worrying about trying to disarm any mines found and simply mark these for the time being and then withdraw and commence a new clearance area next to the one where the mine had been found.

The training programs were delivered at a Pakistani army camp situated at Risalpur; about a fifty-minute drive from Peshawar. The training was delivered in English and then, with the use of Pakistani army staff, was translated into Pashto.

Each contingent had a class of approximately thirty students who were managed and organised by the Pakistani army staff member. This class had a heavy canvas tent to cater for its training and these tents were grouped in an area so each national contingent was training their class in the same area. When the Australian contingent arrived, there were national contingents from USA, UK, New Zealand, Canada, Italy, and Norway. Later we would be joined by a contingent from France. Turkey had also supplied a contingent, but they had returned home and there was no news on a replacement contingent from the Turks.

As the training was done in a hot climate and under these heavy tents, we trained classes in the morning ad stopped about midday due to the heat. Therefore, travel to Risalpur was done in the early (pre-dawn) time and was, to say the least, hair raising. There are very few road rules in Pakistan, or so it would appear, and it's literally every man for himself on the roads. I say man because being a strict Islamic country, women were not permitted out of the household unless accompanied by a male member of that household and definitely not permitted to drive any form of motor vehicle.

To that end also, at this early stage we did not 'self-drive' but relied upon a local driver who was hired by the UN along with his minivan. This was driven at breakneck speed between Peshawar and Risalpur and in the pre-dawn light it was an experience. For a start many vehicles on the road either had their headlights turned off, or were not in working order. As well as cars and buses and trucks, there were people on all sorts of motorbikes and walking and riding donkeys or on a donkey and cart. How the driver saw many of these people, if he did, was a miracle and he seemed to just get around many of them. Flat strap and stop were the only two speeds available. Like my fellow members, I soon became a wide-eyed 'clutcher' being driven around.'

Bob Kudyba, 1st Contingent

'Once we had completed an initial lesson on mines and their types, we then commenced to focus on building up each student's individual skills; this skills base was accomplished by putting the class through a certain demonstration, followed by each individual practising the drills and actions, followed by another assessment of them with less correction. Mostly these actions were accomplished well by the students. Many Afghans were illiterate, or semi-literate, however they were able to overcome this deficiency by watching and copying

Anthony Smith (right) and a Canadian instructor (left) supervise an Afghan student (centre) at Risalpur in 1989.

what was demonstrated. Those that were literate took copious notes and shared these notes with those who could not read after training hours.

One aspect of the training that was challenging was the delivery of subject matter in English and then having it translated to Pashto. In some cases, some members of the class were Dari speakers, or spoke another dialect. In these cases, the instructor would have to wait until what was initially said in English was translated to Pashto, then into Dari and in these other cases then into that language. From the initial translation would come questions. These sometimes became lost in animated discussions with the translator, who would need to be stopped and told to just cut the discussion short. So, if we were doing a purely theoretical lesson a great amount of time was spent just delivering information passing the information from instructor to class. As time went on, due to this laborious to and fro, we worked on curtailing theory lessons to bare minimum and concentrated on practical sessions.'

Bob Kudyba, 1st Contingent

'Another early challenge was that the UN had little initial control over what lesson content was being presented and each national contingent was very much doing their own thing. This tended to create confusion as the various students would meet and exchange information on what their instructors had taught them that day.

Usually mornings were taken up answering a great deal of questions from classes about what they had heard the people in the American group had been learning, or the Italian group and so forth. It soon became apparent that there needed to be standardisation in the training.

Initially the contingent commanders had a meeting about the issue and then decided to form a training coordination group, where one member of each contingent would attend a weekly meeting focusing on getting us all on one sheet of music. The meetings were usually accomplished smoothly.

In some cases, however, this was not the case. Warrant Officer Imants (Monty) Avotins was the Australian representative and one time he arrived back at the Australian house a bit 'hot under the collar'.

'A problem today, Monty?' I asked him.

'Too bloody right, we spent an hour today discussing and debating when a trip wire is considered a 'high trip wire'!'

'Did it get resolved?'

'Well, I told them when it's chest high or more it's a high trip wire. I don't bloody walk around the jungle or anywhere else with a tape measure in my pocket to measure the height of trip wires!'

By then, the clipboard he had been carrying went across the room to smack against the wall and he reached into the fridge and ripped open a can of beer.'

Bob Kudyba, 1st Contingent

'Another aspect of the training that rapidly needed a solution was the need to use examples of land mines and unexploded ordnance as training aids to explain procedures, identify various types of ordnance and conduct various drills on. The problem was there were no inert examples of these mines and explosive ordnance

An Afghan student learning the high tripwire drill at Risalpur in 1989.

only real, live ones that could still function. So, there was no way we could use live ordnance in lessons. It is strictly forbidden in the Australian Defence Force.

So, I then commenced to take trips to the local Smuggler's Bazaar to try and procure examples of ordnance that could be made safe to use in lessons. The local Smuggler's Bazaar was situated in an area known as the 'Tribal Area'; this was outside the precinct of Peshawar and was accessed through a police checkpoint with a boom gate. In this market one could buy anything and everything (more on this in detail later on); there were many stall holders in the bazaar who dealt in anything explosive, landmines, various types of explosive ordnance such as mortars, artillery shells, demolition charges, sticks of gelignite, anything that could explode. So, I started to get to know some of these stall holders and bartered with them for different examples we could use in classes.

The next challenge was making them safe for our classes; fortunately, by then, the French contingent had arrived and one of their members was an Ammunition Technical Officer (ATO), who just happened to have brought all his tools over with him.

This bloke set himself up in a local cemetery, with a table, a vice and his tool bag, as he observed wryly to me one visit, 'If I have an accident, I don't have far to travel for the funeral, Bob.'

Over a couple of beers in the American Club we secured his services to inert our growing collection of land mines and explosive ordnance. This was making the Boss a bit edgy as we had it stored out in a back dunny at the rear of the house. But the stock of land mines and other items of explosive ordnance were growing.

The Ammunition Technical Officer (ATO) would inert the explosive ordnance by destroying any detonators and cutting the item open and extracting any explosive fill inside the item. He'd them seal the item closed again and we'd paint a white cross over the base and top of the item to signify it was free from explosive content.

My fellow team members were also adept at recognising items of ordnance we needed for our training and also contributed to bringing in this, that and something else. Our stock of 'To Be Inerted' items in the backyard dunny was growing by the week, much to the Boss's consternation and the ATO was being shouted countless beers in the American Club for his work taking the explosive out of the items of explosive ordnance. Voila!'

Bob Kudyba, 1st Contingent

Sequence showing Afghan student practicing pulling an anti-tank land mine to counter any potential booby traps at Risalpur in 1992.

'With the lessons sorted and some examples of explosive ordnance and mines inerted, we were on our way; the next thing was the tool used in classes to prod and excavate the ground. Initially, the Pakistan army supplied some kit they had, which consisted of these real long prodders and such. As one wit observed, these things looked like they were last used at El Alamein against Rommel!'

The UN came to the rescue here with their funds and pretty soon a local contractor was engaged to churn out local examples of prodders, which were simply shafts of metal rod, sharpened at one end and stuck onto a piece of really hard wood that looked like dried out Mallee roots.

As the courses progressed, each student was presented with a canvas tool bag containing a couple of prodders, some painted up red triangular mine signs, some mine markers and a couple of exercise books and a cloth bag with what resembled a load of tea towels with drawings of land mines and EO on them for mine awareness classes. I guess this added to a whole industry starting to develop to support this training, churning out prodders, tool bags, these tea towel mine awareness posters and mine signs. So, progress was being made and we were on our way. By the half way mark of our proposed four-month deployment we had a number of classes under our belt, training aids were being acquired and tools being produced for clearance and training and we had a credible training program.

One drawback of this whole idea, however, was that we were not permitted into Afghanistan under any circumstances. This lack of follow up prevented us from seeing if anything meaningful was being achieved. As each new class of Afghan students arrived, our Pakistani army translators became really skilled in recognising those students who tried to come back for a second attempt. As each class graduated, they were given a canvas bag of demining tools and a certificate and the sum of US$360. To these men that was a small fortune, so some attempted to come back around again, by wearing different clothes and altering their appearance, like shaving their heads, etc. At the start of each new class, the Pakistan army translator working with the instructors from each contingent would survey the class and then point out this man and that man to stand up and go report to the control area. 'These men have been here before, Bob, I recognise him and this man and that man!' A torrent of stern Pashto would see the guilty men scamper away quickly. 'Now we can begin the class, please' the army translator would say to me politely.'

Bob Kudyba, 1st Contingent

'We also included in our training such skills as checking for trip wires, using a 'Trip Wire Feeler', this item consisted of a length of suitable gauge fencing wire, which the user gently employed to see if there were trip wires present. Some mines, including the dreaded Soviet POMZ type mines, consisted of a crenulated iron body, resembling a hand grenade, sat on a wooden stake, the mine was actuated by extending a thin, but strong wire from the fuse, sat atop the mine, off to an object such as a tree, or another mine. The usual place for the trip wire was about 15 cm off the ground; designed to catch the unwary person walking through areas. Some guerrilla groups actually got cunning and fixed these types of mines into trees to catch people travelling in vehicles. Hence, the 'High Tripwire' idea.

The students themselves were very practical people who although, in many respects, were poorly educated, they were very adept at looking and copying what actions were demonstrated to them. From the practical work we then graduated the class on to a day on the demolition range, where they were given instruction on basic use of explosives, detonators and how to place charges next to items of EO in order to destroy that item. In discussions with the students, many Afghans confronted with landmines would either try and detonate the item by shooting at it, or try and start a fire near the item to make the heat of the fire cause the item to explode.'

Bob Kudyba, 1st Contingent

A class of Afghan students at graduation at Risalpur in 1989.
On left is George 'Jock' Turner. On the right is Bob Kudyba (standing) and Alan Palmer.

A Soviet PFM 'butterfly' anti-personnel mine.

'The Soviet PFM mine, commonly called the 'Butterfly Mine' was a particularly nasty item, these mines dispensed from containers and floated to the earth scattering as they fell down. They were reasonably light and made entirely of plastic. They were made in various colours such as bright green and soft to touch. Once the mine struck the ground, an inertia fuze would arm the mine. A constant pressure on the soft plastic body would cause the liquid explosive contained inside to function. There was enough explosive power in one mine to cause serious damage to a foot. Children were especially susceptible to this mine as they would pick them up and play with them, when squeezed the cumulative pressure would eventually cause the mine to function. Result was a child without a hand anymore. The liquid explosive contained in these mines was also highly toxic, especially when it was burned and gave off toxic smoke. So, the Afghan practice of gathering the mines and putting them with wood and leaves and burning them was not a healthy option either. These mines could not be disarmed and they needed to be destroyed where they were found.'

Bob Kudyba, 1st Contingent

An Afghan instructor giving a lesson on anti-tank mines at Risalpur in 1992.

'It was decided to steer clear of attempting to teach students how to disarm, or neutralise, landmines for a number of reasons. The first one was each mine required very precise actions to disarm the mine. Another was that without proper training we were concerned about people becoming blasé about the matter and then becoming casualties of their own carelessness. The reference material to disarm the numerous items of land mines and EO was a couple of trunks load of books that would need a decent truck to cart around so for those reasons we decided that destroy in situ; that is, where the item was found, was the safest option.'

Bob Kudyba, 1st Contingent

'With our day on the range, the explosives supplied were by the Pakistan army and this consisted of sticks of plastic explosive wrapped in wax paper. The Pakistan army also supplied the detonating cord and detonators. Each student would first undertake a 'Confidence Charge', that is, under the watchful eye of an instructor each, a squad of students would walk out onto the range, place the stick of explosive down with detonating cord attached, attach a detonator, to which was attached a length of safety fuze. On the command of the Officer in Charge, each student would then light his charge with a match and then on command turn and walk back a set distance and turn and watch for their charge to explode. Sounds easy. What we encountered was not that easy; on most range days, for some reason, it was windy. The correct matches for lighting safety fuze are specially manufactured to burn in windy conditions. The matches we were supplied were normal safety matches, which invariably blew out constantly as the student attempted to set the safety fuze alight. As there was only a limited time to get all the ten charges lit, this chore become a bit of a headache. I don't know if this got resolved down the track with the later contingents.

The demolition range also had a number of dry creeks running through it and it was a usual experience to have a goat or sheep herder move his flock up onto the range area and emerge out of one of these creek beds just as the bigger charges were set to go; to the consternation of many. Despite the presence of sentries and warning flags, these range days also seemed to produce one or two shepherd transgressors during the course of the practice.'

Bob Kudyba, 1st Contingent

'Every morning travelling back from Risalpur we would pass this high stonewall with a painted sign daubed on it: 'Down with UK, USA, USSR, India, China and Israel!' exclaimed the sign-in English. I guess the writer had covered all his bases in the protest movement there.'

Bob Kudyba, 1st Contingent

'As the UN had made this mission a humanitarian mission, the wearing of uniforms was not part of the daily routine. We initially wore the Australian issued 'Jungle Green' uniform without insignia or badges of rank. Once in country we were sent around to a tailor who measured us up for a sand-coloured safari type suit, which was worn with a brassard on either arm, on one side was the UN badge and the other was the 'Operation Salam' badge. I still have these two badges and intend having them mounted up one day.'

Bob Kudyba, 1st Contingent

'Like all good things, they must come to an end. Our four months seemed to fly by and before we knew it, we were getting ready to go back home. A new contingent of Aussies arrived and we were soon immersed in handing over duties to the new team and then travelling around saying our goodbyes. Part of the hand-over was to take the new team out to Risalpur and walk them around the place. They got to meet the other contingents, saw where the training tents were located, the toilet block-where we'd all rushed numerous times with a dose of the 'trots'. They also got to meet the current class of students we had and had to shake hands with the whole 30 of them. As explained, in Afghan culture to not offer to shake hands can result in offence being given. I remember standing off and looking at two teams of Aussies, one in sand-coloured safari suits and the other one in Jungle Greens and

Afghan students practicing laying demolition charges at Risalpur in 1992.

On the demolition range at Risalpur in 1989.

thinking it was great we did come along and try and make a difference to these people's lives. Little did I realise at the time that Australian soldiers would go on to spend a considerable time with the Afghan people. Later contingents would be permitted to make the journey into Afghanistan and actually assist with clearance tasks.'

Bob Kudyba, 1st Contingent

'The US contingent guys were Special Forces not engineers. In late 1989 one of the yanks had an accident when he was disarming a live fuse. His left hand got pretty banged up and he got some fragments in his face too.'

Anonymous, 1st Contingent

'Once settled in Peshawar, the 2nd Contingent got to work. My role was as the Executive Officer (XO) of Demining Headquarters Peshawar, the Operation Salam agency that ran its demining effort in Peshawar (the one in Quetta was run by, not surprisingly, Demining Headquarters Quetta). The Headquarters was a multinational beast, with most of the key 'operational' positions filled by members of the multinational contingents and those of a logistic nature occupied by locally-hired Pakistani and Afghan employees. When I arrived in March 1989 the HQ occupied what seemed to be a large private house in the Rahatabad area, adjacent to the Forest College precinct of the University of Peshawar. This necessitated a commute of about 20 minutes from the contingent housing in Hayatabad. More ominously, it seemed to be close to the Peshawar HQ of the Jamyat e Islami Party, the largest party of the exiled Afghan Interim Government. From time to time, burly Afghan gentlemen wearing uniforms with a latter-day Gilbert and Sullivan appearance would form security cordons in the neighbourhood: we guessed this portended an impending visit by Professor Rabbani himself. When the lease for the Rahatabad building came up in late 1989, it was decided to seek an alternative location in Hayatabad, which was found and occupied by early January 1990.'

Andrew Smith, 2nd Contingent

'During our time, the HQ was led by an 'Officer in Charge' (OIC), an American Lieutenant Colonel who was independent of the U.S. training team, although he was the senior American associated with the program in Peshawar. The OICs did a six-month tour—the third rotated in during our tour. The HQ was organised similarly to a military HQ, with staff sections for Operations, Administration and Logistics. In addition to me as the XO, the Operations and Administration sections were led by multinational officers from the US and French contingents, respectively, although this allocation of responsibility varied as the composition of the multinational presence ebbed and flowed. Locally engaged Pakistani and, in one case, Afghans made up the rest of the workforce, especially in various logistic functions.

Working at the HQ was like stepping back in time, even for 1989—everything was essentially paper-driven, combining the worst of notoriously slow and convoluted UN bureaucracy with local clerical practices that had not evolved since the days of the Raj. I think we had one stand-alone 286 computer for word processing and very basic data management in the Administration Section, everything else was either handwritten or, for things that were particularly important, there were two electric typewriters. One of these was operated by Mafooz the HQ 'Stenographer,' a rather prickly forty-something Pakistani gentleman. The single dodgy telephone line was operated by Asma the telephonist, a Pakistani lady whose fingers were worn down to nubs from repeated rotary dialling of numbers to eventually secure a tenuous connection, which would usually drop out in mid-call and need to be re-dialled. All of this frustration was offset by a kitchen, staffed by a couple of 'cooks', who would provide tea (usually) or instant coffee on demand.

One task that sums up the 'mandraulic' nature of work in the HQ while providing insights into the Afghan refugees who were the core purpose of the mission, was the production of Course Certificates for graduates of the demining courses delivered by the training teams out at Risalpur. This was done by the Administration Section every three weeks, to match the cycle of courses. The process began at Risalpur with the collection of identity information—name, father's name and village—and a polaroid photo of each subject. These were key identifiers in a country where so many men had the same name. This information was manually compiled into long typed lists that were brought into the HQ where a clerk would again type each individuals' details onto a printed certificate card.

When the cards were typed, the whole batch of about 500 was sent out to Risalpur to be signed by the recipients. This provided an interesting insight into the Afghans—about 50 per cent of the cards were returned with a thumbprint instead of a signature, indicating that the subject was illiterate. When the 'signed' cards were returned, a polaroid photograph of the subject had to be matched to each card and then glued on in the appropriate space. During this labour-intensive stage of production, a visitor to the HQ could walk from office to office to see staff members deemed to have 'spare time' gluing photos onto their allocated bundle of cards. Once the photos were added, certificates were laminated two at a time in another manually-loaded machine, then individually cut and trimmed from the lamination sheets with scissors. After a final quality control check, the completed run was sent out to Risalpur in time for the graduation ceremony. The Administration Section could then devote itself to other tasks for a while, until the entire process began again.'

Andrew Smith, 2nd Contingent

Andrew Smith and his Pakistani driver at Risalpur in 1990.

'Operation Salam differed from classical 'peacekeeping' missions, in that it was strictly humanitarian in nature and operated under the General Secretariat of the UN, rather than the Security Council. This affected us in a couple of ways. One was a conscious effort to downplay the military side of the demining program by controlling the optics of the mission. For example, the person running Demining HQ Peshawar was deliberately styled the 'Office in Charge' (OIC), not the 'Commander.' The most obvious optic, however, was what we wore. Although a uniform was conceded as necessary to distinguish the instructors at the training camp and to facilitate relations with our partners the Pakistani Army, that uniform needed to be 'neutral' and not overly 'militaristic.' This was solved by devising a common khaki-coloured 'safari suit' style ensemble that was, even in 1989, well and truly out of fashion. National origins were distinguished by brassards bearing national flags or badges as well as the UN and Operation Salam logos. These uniforms were made by Mr Farooq, a local tailor with a shop in the Sadar market district who would appear on his motor scooter, measuring tape and order book in hand, as each incoming contingent arrived in Peshawar. Production of the uniforms took a couple of weeks, during which time newcomers would usually wear 'demilitarised' national gear—in our case, this was the Australian green fatigue uniform that was then being phased out, sans badges of rank and with trousers unbloused. No-one was particularly comfortable with this and we were very glad when Mr. Farooq eventually delivered our new clothes.

The humanitarian nature of the mission also affected us through the professional and social environment we found ourselves in. The international effort to support the Afghans—both refugees in Pakistan and those still struggling across the border—was massive. In addition to a range of UN agencies and the Red Cross, we were introduced to the non-government organisations (NGOs) that make up much of the global 'aid industry.' There were too many of these to count, each with its own peculiarities. Somewhat irreverently, we came to refer to NGO members as 'do-gooders.' This was not really a cynical classification so much as a recognition that, legitimately, most of them were seeking to make a positive contribution to a real human problem.

As well as being a UN agency, the Demining HQ had support contracts with a number of NGOs. These two aspects brought those of us who worked there into professional contact with do-gooders from time to time. As a contingent, our most frequent interaction with NGOs was at the watering hole to which we were all drawn—the US Government Employees Association Club in University Town. Usually known simply as 'the American Club,' this was one of only a couple of places in Peshawar that served alcohol, so it had an enormous attraction for those who liked a drink regardless of their affiliation. The do-gooders were an incredibly eclectic bunch, from most Western nations and all walks of life, and we got to know many of them well. Many were impressively accomplished people, but more than a few were also on the rebound from a failed relationship or some other personal catastrophe. It was almost as if service in an NGO had become the late twentieth century alternative to running away to join the Foreign Legion. A few were 'peacenik' ideologues who were stand-offish to we 'deminers' because of our military connection, but all in all the do-gooders added a unique and very positive dimension to our experience in Peshawar.'

Andrew Smith, 2nd Contingent

Afghan instructor supervising Afghan students setting up a metal detector at Risalpur in 1992.

'It was disappointing to find on graduation days that the small canvas bags with the mine clearance tools we issued the students were being hawked in the bazaar within hours.'

Anonymous, 2nd Contingent

'The training aids for mines and munitions that we used, except for a box full of wooden replicas that barely went close to representing what they were intended to be and therefore kept locked away and never used, were scavenged from the local foundry. Pakistani 'jingle' trucks would travel back and forth in and out of Afghanistan smuggling all sorts of cargo in and returning with tonnes of scrap metal. A large portion of the load each time would be in the form of unexploded ordnance.

There was a huge open smelter at the foundry with an open top and the workers would barrow the scrap up to it and tip the contents into the glowing smelter to be melted down and turned into reinforcement bars for concrete construction. There was a mass of large holes in the roof above the smelter from the UXO items exploding as they were tipped into the white-hot liquid metal.

On Friday mornings a couple of us would visit the foundry and walk around and identify and separate out the live UXO from the scrap. We would grab what we needed to use for training aids and disappear with our booty.

The manager of the foundry approached me on one of our Friday missions there and inquired with me that we were making use of the UXO that we were taking with us each time and perhaps he should be charging us for the items. I said that would be ok as long as he paid for our services for separating out the UXO from the scrap to save his smelter. He asked how much we would charge for the service and my reply was that it would always be more than he would charge us for the items that we wanted to take with us. He said that I was a very good businessman and that we would continue with the arrangement that we already had in place.'

Allan Mansell, 2nd Contingent

'The training was conducted at the Pakistan Air Force Academy at Asghar Khan, Risalpur. Where enrolments of 500 Afghan refugees that were residing in make do camps in and around Peshawar would participate in a course that from memory lasted about four weeks. The 500 participants were segregated into 10 groups of 50 and the training was conducted inside tents that were located on a patch of land at one end of the PAF Academy runway along some addition land that served as a closed training area. Across the road and at the base of a small hill where the local cemetery was established, was a site designated as the demolition range for live firing of explosive charges.

The participants in each course were paid a per diem to attend, that was intended to cover costs for their travel to and from the training, meals, etc. The changeover in contingents coincided with the commencement of a new course and we newbies were introduced to the day one confirmation and culling process. It was identified during earlier courses that there were quite a few familiar faces among the participants. This phenomenon was a by-product of paying the per diem. Many who had already qualified on previous courses would shave their beards off, or grow one, shave their heads and attempt to alter their appearance in other ways. They would then enrol under an alias and try to collect another four weeks of per diem.

The method introduced to us by our predecessors was to take class photos on day one and then slip away and check the faces in that group photo against class photos from previous courses and identify as many of those that were doubling up as possible. Needless to say, the system was not that effective, but a fair few identified all the same and would be replaced by another volunteer. There was no shortage of volunteers of course, given that the Afghan refugee population in Peshawar at the time numbered 3 million.'

Allan Mansell, 2nd Contingent

'As previously mentioned, there were 500 participants allocated to each basic demining course and they would be divided into groups of fifty individuals. The primary ethnic composition of Afghan people consists of Pashtuns, Hazaras, Uzbeks and Tajiks and the languages most commonly spoken are Pashtun, Dari/Farsi and many who participated in the training also spoke Urdu, the language most commonly spoken in Pakistan.

Each class sized group of fifty would be a mix of all ethnicities and that meant a requirement for more than one interpreter per class and often saying good morning could take 15 minutes or more at the beginning of the day's training, let alone the length of time that it could take to explain and teach anything else. That was where the language training that we received from ASC1 as limited as it was, became very valuable and saved quite a lot of time in getting training across. The interpreters would monitor what we were saying too, in order to ensure that we were actually saying what we thought we were saying.

As a bit of side note regarding the topic of the limited language training, when the replacements arrived for the departing Canadian instructors arrived, they were moving around from class to class as part of their induction. They happened visited the Aussie classes while we were conducting some practical testing on clearance techniques. They asked later on in the day, how long the language courses were that we obviously had done in Australia prior to being deployed to Pakistan.'

Allan Mansell, 2nd Contingent

'Demolition range day was always a mixture of mastery and mayhem. After the Afghan trainees had been trained in explosives safety, theory and practical preparation of small explosive charges and placement for the destruction on landmines and UXO, they would be subjected to a day on the demolition range preparing, handling and detonating explosive charges.

The morning for us trainers would commence with a drop in at the Old Bazaar in Peshawar on our way out to Risalpur, to collect a few decent sized watermelons. We would arrive

Afghan instructors giving a lesson to Afghan students at Risalpur in 1992.

at Risalpur and shake out our 150 trainees from the Aussie classes and divide them into details of ten.

We would then marshal them all up onto a nice viewing point where we would go about setting up a non-electric initiation set, place it into a watermelon and detonate it to demonstrate the effect of a single blasting cap. After the detonation chunks of watermelon would be scattered over about a 5-metre radius and there would be a 150 Afghans above all shouting Allah Akbar at the tops of their voices.

The next serial would be to put the 15 details of 10 personnel through preparation of the initiation sets, then they would move through and collect a half stick of Pakistani manufactured PE3 and move onto the confidence range to set up and detonate their explosive charges.

I had the misfortune of working the initiation set preparation location on one range day and without an interpreter within cooee. I was totally reliant upon my mastery of the four different languages that were native to the trainees. Unfortunately, I only knew the Pashtun word for stop, which is 'bus'.

Now the way that we Aussie Engineers crimp our detonators onto the Fuse Blasting Time (FBT) to a blasting cap or detonator they are one and the same, to form an initiation set is to cut the FBT to the required length, take up the FBT in one hand and a non-electric detonator in the other hand, check the fuse well in the detonator is nice and clean and clear of any foreign matter. Then the FBT is slid into the detonator and the two components are clasped together nice and firmly with the thumb and fingers of one hand. Next, we take up a pair of specialised crimpers, place the jaws around the section of the detonator containing the FBT and close the jaws firmly enough to ensure that the crimpers will not slide down the detonator. We then reach around with the initiation set and crimpers into the region of our butt cheek and once there, turn the head and eyes away and perform a nice firm crimping of the detonator onto the FBT without over crimping and risk severing the powder train inside the FBT. The initiation set is then checked to ensure that it is all assembled properly and if so, the initiation set is held in one hand with the joined section between the forefinger and thumb for carriage. This method is employed so that if there were to be an accidental detonation, the face and eyes don't get caught up in the blast and the damage is limited to hands and butt cheek.

However, that was not the method that was required laid down in the SOPs for training at Risalpur. The crimpers to begin with had twin jaws. One with a sharp scissor edge and the forward jaws were the crimping section. The purpose of the design being that the scissor jaws could be used for cutting FBT and detonating cord and the crimping section for assembling initiation sets. That sounds great in theory of course. Except of course when the ignition set finds its way into the scissor jaws and things get quite pear shaped and you have to point at things using your nose instead of your fingers from that time on.

The method required under the SOPs was to assemble and crimp the detonator to the FBT at eye level. The method in theory was intended to ensure that you could clearly everything that you are doing all of the time and avoid mistaking one set of jaws from the other while retaining the FBT firmly inside the detonator at all times. A fairly major flaw in the latter method is that they did not take into account 150 very nervous individuals preparing their own initiation sets for the first time.

The trainees would arrive at my location one at a time and I would point out to them where everything was located. Hand them a sharp knife and show them how and where to cut the FBT. They were then instructed to curl the FBT up nicely and hold it in one hand while I handed them a detonator that I had first checked. It is at about this point that their adrenalin metre would climb to 75%. They were then talked through the procedure to get the ignition set pressed together and up in front of our faces ready for crimping. They would then be given a demonstration by me placing the crimpers onto the initiation set in the correct position and closed against the detonator. I would then remove the crimpers and hand them to the trainee to mimic what I had done. At this point their adrenalin meter would increase to warp factor five and their hands would be shaking and the scissor jaws would go straight around the detonator with me shouting bus, bus, bus and getting my fingers in amongst crimper jaws eventually forcing the mayhem to a halt and repeating the same torture until they settled down enough to get it done correctly. By the end of the day my hands had cuts and skin missing all over the place and I was bleeding like a stuck pig. However, we all got the procedure down pat in the end.

Setting up and lighting up the confidence charges certainly had its moments also. The fuses on the initiation sets were cut so that there were thirty second intervals from one fuse to the next for the ten charges. The Pakistani matches that were available were notoriously unreliable and the old brown FBT could also be inconsistent and burned at 17 seconds per foot, which is approximately double the burning rate range of 36-47 seconds that we would normally apply when using our own service FBT.

The ten trainees in a detail would be required to shout 'Infajar' in four different directions, which is a Pashtun term that roughly translates in English as 'explosion now' and is understood all over Afghanistan, and then they would light up one at a time longest fuse to shortest until all fuses were lit. They would be required to stand up and cover off their charges once they were lit and once they were all going, they would be instructed to file of the range and move to the save area, turn and watch their charges detonate. We always made sure that we carried plenty of spare matches as between the crappy matches and the sweaty hands of the very nervous trainees we often got down to the minimum time left to move back.

The trainees were taught to place a hand either side of their mouths when shouting 'Infajar' to make their voices project further in a certain direction. However, when they pray, they place their hands cupped behind their ears and that is what they would do each time that they shouted. That adaption led to other variations on the theme, that were implemented by the Kiwi trainers where the trainees were taught to stand on one leg and dance around in a circle shouting infajar. There may have been a few other adaptions to the infajar dance implemented from course to course by the trainers from the various international contingents, but I am just going to leave it at that for now.'

Allan Mansell, 2nd Contingent

'Word got around regarding the fact we were military landmine and UXO experts and that led to some interesting items being deposited with the chowkidars (house security guards) from time to time. One such item was a P40 rifle grenade UXO. The P40 rocket grenades are normally fired from an RPG7. So SSGT Ian Mahoney and I decided that we might take it out to the Demolition Range at Risalpur next time that someone is on the range and do a bit of experimenting with it. The aim of the experiment was to attempt a low order technique to see if the venturis and the base of the rocket could be removed using det cord, without detonating or high ordering the rocket.

The opportunity that we were awaiting presented itself when the Kiwi trainers were putting their trainees through on the Demolition Range. Ian and I had a chat with Captain Chris Fauls the New Zealand Contingent Commander who was in charge of the Range Practice on the day and explained what we wished to do with the item but out of the way from their activities and between their serials. He agreed and away we went. We found an ideal location to conduct our little experiment in a natural reasonably wide and semi enclosed ditch. We set up the item and cranked off the det cord, waited the appropriate soak time and the moved back to it to check the results. The experiment had all but removed the rear section from the rocket grenade with detonating it.

We went back up and spoke to Captain Fauls about giving it another go, but by that time he wasn't very keen to continue as he had concerns about dealing with UXO on the Dems Range. I pointed out that the SOPs allowed for the destruction of UXO on the range. He interpreted the SOP referring to UXO that were discovered on the range, not UXO that were brought to the range. That was a fair enough call, so we agreed to just high order it and leave the range.

So back we go to crank the item off. The rocket grenade was already nose first into some soft mud (Hmmm mud?? There will be more on that in a moment), so we set up the donor charge on the item to attempt to make it function through its normal firing principal to fire the shaped charge, just to see how much penetration into the clay and rock beneath it would achieve. The result would provide some useful knowledge for explosive earthworks sometime down the track. We went through with the detonation and retuned to discover that it had indeed fired the shaped charge and broken up the dry clay and rock below for about 2m down and about a metre to a metre and a half across.

The other discovery that we could not avoid noticing as we advanced towards the location of the demolition of the UXO, was the pungent putrid aroma of sewage. It turned out that the site was not just a perfect spot out of the way, out of sight and protected for experimenting with low order techniques on UXO, it was a very popular shit pit for a whole lot of Afghans.

So that was another aspect of the experiment that the Kiwi officer was not too pleased with us about, as we made good our escape and they were left to continue with the rest of the practice.'

Allan Mansell, 2nd Contingent

'I had been assigned a small, old and rundown Toyota Corolla sedan by ATC as my personal vehicle for everyday use. This was the car in which my driver, Siddiqullah, had collected Judy and me from the Marriot hotel in Islamabad way back in January. It was at least ten years old when I was given control of it, so you can imagine what sort of condition it was in. It had several dents and buckles to the doors and most of the body panels and almost every joint rattled. The tyres were near to being bald and the headlights were as powerful as a second-hand torch. Yet, the vehicle did have a surprisingly good feature: it blended in with the local community of cars. No one would have ever thought that a western foreigner would be driving such a vehicle. Typically, the fancy Toyota Land Cruiser or Nissan Patrol were owned or used by international folk at the time, but in the far reaches of Pakistan, following an unpopular Gulf war, this was exactly the type of car I wanted. In fact, I made some improvements to it., with the most prominent being large stickers and posters placed across the rear window displaying Saddam Hussein and his military might. I left these on the car throughout the year. There were pictures of fighter jet planes and tanks and rockets and soldiers with guns, all firing into the air, with Saddam's big head in the centre. I also took to wearing a patu when in the car after work. A patu is really just a thin blanket that is a traditional and common garment worn throughout the region. In this manner we had no problems when driving around.'

Graeme Membrey, Technical Advisor

'Despite the fairly obvious requirement to maintain a low profile, several foreigners we met and knew well, refused to do any such thing. Unfortunately, we knew of one Dutchman and his Belgian wife who were badly beaten just outside of Peshawar in late February when the husband was replacing a flat tyre. They were beside the road on the outskirts of a nearby large village. As he was fitting the new tyre, two car loads of men pulled up and started an argument about their nationalities, background and religion. The Belgian woman locked herself in the car, but got out to help when she saw her husband knocked to the ground and beaten with the wheel nut spanner he had been using. She was not raped, but was physically assaulted and groped by the men as other cars apparently drove past with scant regard. The husband was badly injured and spent several days at the international hospital in Islamabad.'

Graeme Membrey, Technical Advisor

'In Peshawar there was a facility open to international people that no Pakistanis (or Afghans for that matter) was permitted by law to enter unless employed by the management. It was the American Club: an entertainment club that, of course, had a huge bar serving Jack Daniels, vodka or beer as was desired. As you would imagine, it was attended by a broad swathe of interesting characters. One night while I attended a relaxing evening there, two patrons who I knew quite well, started an argument that seemed to go on and off, over the three hours or so that I was there. One was an Irishman and the other a Welsh-Australian. In the end, they took their argument outside when the club closed and before any fisticuffs took place, both of them angrily drew pistols and began threatening each other. As I was right there, in a reflex moment, I grabbed the gun from the Irishman and two colleagues grabbed the other man's gun so that no shooting would go ahead. It took some time and much effort before they both settled down and decided not to shoot each other—at least not then.'

Graeme Membrey, Technical Advisor

Afghan instructor giving lesson on first aid to Afghan students at Risalpur in 1992.

'At one point we thought it might be a good idea to do a 'land mine buyback program'. That is, to purchase the mines and take them out of circulation. This was because a lot of mines were being recovered and either re-used to protect crops and livestock, or being sold into the bazaars. We heard later though that by putting a decent value on the mines that we had inadvertently created a market and that children were being told to go out and recover the. The land mine buyback program was stopped pretty quickly after that.'

Anonymous, 4th Contingent

'As the Christmas of 1990 approached the US Special Forces team began to thin out. The formal version, which was short on detail and long on un-answered questions, was that the troops were heading home for Christmas leave and would be returning sometime in 1991. As part of the thinning out, the United Nations overall Commander in Peshawar—a US Army Military Police Lieutenant Colonel (who never appeared at Demining Headquarters) also departed. With the two NZ Officers operating more or less independently in Quetta and the resultant draw down on US forces, our contingent was left to fill the gaps. An informal chat over a few beers with a sympathetic staff member of the US Embassy revealed that the US Special Forces contingent would be taking their 'Christmas leave' in Kuwait preparing for their role in Desert Storm. Although the members of 5th Contingent developed good relations with the US team on a social level there was some concern within the contingent about the training program instituted by the US team. The US team's departure resulted in a general reorganisation with me becoming the local commander, Captain Mike Kavanagh assuming the Executive Officer role and Warrant Office Class One Dave Edwards stepping into the Training Officer appointment. We also had additional roles as Technical Advisors to three of the Demining Non-Government Organisations in Peshawar.'

Brian O'Connell, 5th Contingent

'The escalation from Desert Shield to Desert Storm impacted significantly on the way the team lived and recreated in Peshawar. Whilst the Pakistan Government was a member of the 'willing', the denizens of Peshawar were rooting for Saddam Hussein. Prior to the initiatives in Kuwait, the locals viewed us something of a curiosity. The pale skins and western clothing would generally attract a long quizzical stare and nothing more. That changed dramatically. Peshawar became the scene of daily riots in which vehicles unfortunate enough to be caught up were generally turned over and set alight. Most nights there were explosions as rival political groups attempted to injure each other and any unfortunate in the vicinity. Europeans were met with hostility. Our house received visits from police (real or phoney) demanding bribes and on one occasion an AK-47 barrel appeared over the high wall of our team house but no shots were fired. Movement around Peshawar became difficult and caution was required before venturing out. The local UN groups attempted to set up a local safety net using hand held radios but this was mostly ineffective as was the marginal local telephone capability.

The role of the team was to supervise training at Risalpur which was about an hour and a half down the road from Peshawar. So, to limit visibility the forward journeys were started much earlier in the morning under cover of darkness and the return trip completed in the early afternoon before the university students hit the streets. The Commandant at Risalpur, a Pakistan Army Engineer Colonel, (Colonel Sajad) who was very supportive, provided the contingent with two young Pakistani Army Captains who travelled with the contingent as protection. These two young men were quickly adopted by us and received at the house with open arms and treated to some good Aussie hospitality—enough said. I took the decision that should the team's position become untenable and we needed to depart at short notice, each team member would keep a travel pack with fresh water and food. It was to be kept handy and an alternate route (other than the Grand Trunk Road) to the Australian Embassy in Islamabad was plotted. The Commandant at Risalpur also offered accommodation at the Camp and this offer was kept as Plan B.

One of the difficulties of looking Caucasian was that locals tended to assume that we were Americans, even though our brassards had an Australian flag on one shoulder. In a great example of the humour and the initiative typical of Aussie soldiers, during one tense moment Corporal Dean Brown (Brown Dog), taking advantage of the Pakistani's love of cricket said to a hostile, 'Do you know Allan Border?' 'Well we're from the same village.' This instantly defused the issue. Brownie had a quick wit and a great sense of humour which contributed to team moral and was very handy on this occasion.

Once Desert Storm commenced, 'flying coaches' and vehicles of all sorts plastered their windows with 'Saddam Posters'. These were cheap prints showing the head or full body photos of the man. The team drivers wanted to follow the fad and it was agreed, as the photos became a form of camouflage for the white-faced vehicle occupants. Desert Storm came to a successful conclusion in February 1991 and by the time 6 ASC arrived the bombings and riots in Peshawar had almost subsided and life was returning to normal.'

Brian O'Connell, 5th Contingent

An Afghan student learning how to use a mine detector at Risalpur in 1992.

'For some reason I was the only one that went to Risalpur to work on this particular day. From memory it was a Saturday and I was coming home in the van driven by Ramat and we were coming through the old city portion of Peshawar. I was tired so decided to lie on the back seat of the van to catch some sleep. Happily dozing I was suddenly bought back to life with an explosion and shower of glass from a side window. Thinking the worst, I look up and there it was. A horse's head protruding through the window staring at me. We had been 'T' boned by a horse and cart. I sat there waiting for this thing to suddenly yell out, 'Wilburrrrrrr' or the Pashtu equivalent.'

Ben White, 5th Contingent

'I remember the Russians donated a heap of gear most of which was not much chop which sat in a room at the ATC office in Hayatabad. There was a bucket load of EMP Mine Detectors, prodders, etc. Among them was many hard-cover purple books which were written in Russian, so they were quite useful (tongue in cheek). I still have one and interestingly there are numerous diagrams in them some of which depict the various methods that the Mujahideen used to deploy mines and improvised explosive devices against the Russians. A decade or so later Australians would face the same devices and methods this time being used by the Taliban.'

Ben White, 5th Contingent

'There was a mud brick house at Risalpur that was used for training deminers in building clearance. I can't remember how it came about, but we came up with an idea to wire the house so that we could put electrically initiated switches in various rooms remoted off to a 'live' main charge at a safe distance away. The idea was if someone made a mistake then it would initiate the charge. Great idea. I remember spending hours digging small trenches and camouflaging twin core cable into the wall of the building with mud. All done we did a test charge, packed-up and returned home. The next day we had an activity where we would get to test the new concept. We arrived early ready to set-up so we head to the wires to start connecting our devices. 'Where the fuck is the wire?' Jamal one of the Afghan trainers who was with us points toward a group of nomads that had encamped close by explaining how they likely harvested the wire for the copper. They also used the house to shit in as there were several fresh 'surface lays' all over the house. Needless to say, we never did that again.'

Ben White, 5th Contingent

'The Administration Officer role on Demining Headquarters was pretty boring so, fortunately, I had some other roles including Technical Advisor to OMA, Technical Advisor to MCPA, Contingent 2IC and later, when the Americans left to go to the Gulf War, Operations Officer on Demining Headquarters. I also got to go to Risalpur to help out with training, usually as the OIC for the Demolitions Range Practices.

This was a lot of fun as it was based on the Confidence Practice used in Australia with each trainee setting off a stick of Pakistani explosive. They would prepare their charge, place it down and call out 'Infajar' which was Urdu for 'Fire in the Hole'. They would do this in four directions so everyone knew there was about to be an explosion which was good as the range was just a paddock and while we tried to keep people away, some were just keen on taking a short cut. When the command to 'fire on' was given, they would pull out their own matches which is a safety breech and the Safety Officers would be smacking them away and stomping on the matches. The Afghans refused to hand in their matches at the start of the practice as they were relatively expensive and they didn't expect them back at the end of the practice. When the command was given to move away from the explosives, they would move quite quickly to the safe area but then start to move back towards the explosives so they could be closest when they went off in a show of bravado.

On 2 August 1990, Saddam Hussein ordered the invasion of Kuwait resulting in a US led coalition force being deployed to Saudi Arabia quite quickly in response. It was against this background that we conducted our pre-deployment training and first few months in Peshawar. Strangely, I don't remember this being a key issue in our security brief before our departure.

In November 1991, the UN Security Council set 15 January 1991 as deadline for Iraqi withdrawal from Kuwait. Iraq failed to comply and on 17 January 1991, coalition forces began an air bombardment of Iraqi targets in Kuwait. On 24 February 1991, the ground war began and the First Gulf War was over shortly after that.

The Mine Clearance Program would run monthly meetings in Islamabad and I believe one was scheduled for 17 January 1991 or a day or two before this. These were a good opportunity to meet with the Islamabad and Quetta based people involved in demining as well as a chance to unwind in Islamabad.

Brown Dog and I arrived at the Islamabad offices the afternoon before the meeting, having driven down from Peshawar, and reported to Lieutenant Colonel Selwyn Heaton who was a Royal New Zealand Engineer officer. He asked us what we were doing in Islamabad and we advised that we were there for the monthly meeting. Selwyn advised that it had been cancelled due to imminent commencement of hostilities in Kuwait and the UN wasn't sure how the Muslim world was going to respond. We had obviously missed that memo. We had been booked into the Pearl Continental Hotel but Selwyn advised this was too dangerous so we would be staying at his house until arrangements could be made to return to Peshawar.

We ended up staying with Selwyn for about six days watching events unfold on CNN. The air war was going quite well for the coalition but Saddam Hussein was full of bravado with his 'Mother of all Battles' rhetoric. There were a few other UN people from Peshawar who were stuck in Islamabad so a convoy was arranged to return us all to Peshawar under the command of Colonel Sajjad, a Pakistani Army Engineer Officer who was based at Risalpur. We had met Colonel Sajjad a few times and he was an impressive officer, tall with a big British Army style moustache and a swagger stick which he carried everywhere.

The convoy comprising about 15 or so vehicles formed up with us about five cars back in our Toyota Corolla and we headed out of Islamabad for the Grand Trunk Road. Islamabad is about 10 km off he Grand Trunk Road so that part went without incident but we had only been on the Grand Trunk Road for a few minutes when we came across a large demonstration where the road had been blocked by three or four buses parked across the road and there were a few hundred demonstrators there, including quite a lot on top of the buses.

I remember thinking that if the first few kilometres were like this, what was the next 150 km going to be like and perhaps we should turn around and head back to Islamabad but Colonel Sajjad had other ideas. From our car, we could see Colonel Sajjad sticking his head and arm out the window and he was waving his swagger stick at the crowd telling them to clear the road and sure enough, the people on top of the buses got down, the buses backed off the road and it was cleared enough to let our convoy through. If only the military had this authority in Australia. We drove through the demonstration and, after the last vehicle was through, the buses went back into position and the demonstration picked up where it had left off. Colonel Sajjad hadn't even needed to get out of his car.

There rest of the journey to Risalpur went without major incident when Colonel Sajjad's vehicle turned off the Grand Trunk Road and headed into Risalpur. We weren't sure what was happening as the rest of the convoy kept going to Peshawar so we stayed with the convoy until the traffic in Peshawar made it too difficult to stay together, eventually getting home OK. We caught up with Colonel Sajjad a few weeks later and he advised that he went into Risalpur to drop some people off before planning on escorting us through Peshawar as he felt this was the area most likely to have demonstrations. Fortunately, we made our way back to Hayatabad without incident.

We also received quite a few death threats around this time. The death threat would be scrawled onto a piece of paper, usually in pencil, tied to a rock and thrown over the wall into our yard. The guys instructing at Risalpur were up early and would collect the death threats and put them on the notice board in the kitchen. Some days there would be two or three additions. They were all very similar in that they were on small, dirty pieces of paper and said pretty much the same things about the USA being the Great Satan and the UN being the Lesser Satan and to go back where we came from or they would send us to Hell and were generally signed 'From the Mother of All Battles'.

We didn't take them too seriously and I don't think we even kept them but, looking back, perhaps we should have given the proliferation of weapons around the place. We did, however, prepare for a quick evacuation by putting a 'grab bag' of essentials in our Alice Packs that we could pick up and go within a couple of minutes, if required. The plan was that if things got too hot in Peshawar, we would make our way to the Australian High Commission in Islamabad until help arrived. We also rehearsed going a back way out of Peshawar in the mini bus but, once out of Peshawar, there was only the Grand Trunk Road to Islamabad which would be difficult to traverse without incident.

The House Staff were quite supportive during this time and offered to shelter us in their villages. My driver Numnar said we could come to his village and we would be safe as they had mortars and heavy machine guns. I'm not sure if this was true or not but as there was a heavy machine gun set up on top of the Afghan Interim Government House about four doors down from our house, complete with sand bag protection, Numnar's claims were probably correct. It was also another example of how supportive the local people were of these visiting Australians.

Fortunately, the air war was going well and support for Iraq within Pakistan fell off quite quickly, even before the ground war commenced.

As the Operations Officer on Demining headquarters we would monitor mine clearance operations in Afghanistan. If there was an incident, the SOP was to cease mine clearance for that day to allow the team members to recover mentally before recommencing. I remember one day that we were advised that a mine clearer was killed when a device functioned and the call went out to surrounding teams to stop work. We were then advised not long later that a second fatality had occurred when a member of the team was walking along a small stone wall to tell his team mates of the accident and to stop clearance work when he slipped off the wall onto the uncleared side and stepped on a mine.

While we were working with the clearance teams when they rotated through Peshawar, these fatalities reinforced why we needed to be with the teams in Afghanistan to improve supervision and safety. The 4th Contingent had identified this need and made a submission to Canberra to be allowed to go into Afghanistan. Canberra thought about it for a while and sent back quite a few questions to our team that needed to be addressed before they would allow Australians into Afghanistan. The main one was casualty evacuation as the nearest decent hospital was in Singapore. I believe Bear O'Connell provided acceptable answers to these questions as we received approval to enter Afghanistan about two weeks before the 6th Contingent was due to arrive.

We knew this wasn't much time but, with this approval in place, we headed down to get approval from the Pakistani authorities to enter Afghanistan. Unfortunately, we were advised that it takes four weeks for the approvals so we started the paperwork for the 6th Contingent. I believe that it actually took the 6th Contingent a few months rather than four weeks before they were finally able to set foot in Afghanistan.'

Michael Kavanagh, 5th Contingent

Barry Veltmeyer discussing demining drills with Afghan instructors at Peshawar in 1992.

'Very early on in my tour the Chief of the General Staff for the Australian Army came to Peshawar to visit us and get a sense of our work. We put on a demonstration for him and then he took the opportunity to ask some questions. As he was asking me a question one of our Afghan instructors came up beside me and held my hand. In Afghanistan this gesture reflect friendship and respect and he was hoping to communicate to my big boss that we were a good team and working well together. Unfortunately, not being attuned to local Afghan customs the Chief of the General Staff glanced downwards and developed a slightly puzzled look on his face. I smiled but he never broke stride and went on with the tour. I later asked his aide to explain to the Chief what that was all about...I hope he did pass it on.'

Marcus Fielding, 8th Contingent

'In late April we were all going to go to Islamabad for Anzac Day but Mark O'Shannessy got a really bad dose of the shits and I volunteered to stay behind and make sure he was going to be OK. On Anzac Day itself we heard that Kabul had been captured by the mujahideen. It was another night of wild small arms firing into the sky. We stayed indoors. We had to pause the trips into Afghanistan for a few weeks after that to work out how the security situation was changing.'

Marcus Fielding, 8th Contingent

Afghanistan

Afghanistan is a landlocked country located within South Asia and Central Asia. Afghanistan is bordered by Pakistan in the south and east; Iran in the west; Turkmenistan, Uzbekistan, and Tajikistan in the north; and in the far northeast, China.

Its territory covers 652,000 square kilometres and much of it is covered by the Hindu Kush mountain range, which experience very cold winters. The north consists of fertile plains, whilst the south-west consists of deserts where temperatures can get very hot in summers. Kabul serves as the capital and its largest city.

Human habitation in Afghanistan dates back to the Middle Palaeolithic Era, and the country's strategic location along the Silk Road connected it to the cultures of the Middle East and other parts of Asia. The land has historically been home to various peoples and has witnessed numerous military campaigns, including those by Alexander the Great, Mauryas, Muslim Arabs, Mongols, British, Soviet, and since 2001 by the United States with NATO-allied countries. It has been called 'unconquerable' and nicknamed the 'graveyard of empires'. The land also served as the source from which the Kushans, Hephthalites, Samanids, Saffarids, Ghaznavids, Ghorids, Khaljis, Mughals, Hotaks, Durranis, and others have risen to form major empires.

The political history of the modern state of Afghanistan began with the Hotak and Durrani dynasties in the 18th century. In the late 19th century, Afghanistan became a buffer state in the 'Great Game' between British India and the Russian Empire. Its border with British India, the Durand Line, was formed in 1893 but it is not recognized by the Afghan government and it has led to strained relations with Pakistan since the latter's independence in 1947.

Following the Third Anglo-Afghan War in 1919 the country was free of foreign influence, eventually becoming a monarchy under King Amanullah, and later for 40 years under Zahir Shah. In the late 1970s, Afghanistan in a series of coups first became a socialist state and then a Soviet Union protectorate. This evoked the Soviet-Afghan War in the 1980s against rebels.

By 1996 most of Afghanistan was captured by the fundamentalist Islamic group the Taliban, who ruled most of the country as a totalitarian regime for almost five years. The Taliban were forcibly removed by the US and NATO-led coalition, and a new democratically-elected government political structure was formed.

Afghanistan is presently a unitary presidential Islamic republic with a population of 35 million, mostly composed of ethnic Pashtuns, Tajiks, Hazaras and Uzbeks. Afghanistan's economy is the world's 108th largest, with a GDP of $64.08 billion; the country fares much worse in terms of per-capita GDP (PPP), ranking 167th out of 186 countries in a 2016 report from the International Monetary Fund.

Soviet Mine Warfare

The Soviet and DRA military forces in Afghanistan laid a significant number of land mines—probably in the order of several hundred thousand—during the Soviet-Afghan War 1979-1989.

The troops placed anti-personnel mines around their security posts, military bases and strategic points for protection; in the outskirts of cities to prevent the approach of mujahideen forces; as well as in and around villages to depopulate them to reduce local support for the mujahideen. Sometimes the mines were booby trapped to prevent tampering.

Often these mine fields were fenced and marked but over time these fencing materials were scavenged for other uses. In conformance with Soviet military doctrine the details of these minefields were also recorded with boundaries and the number of mines laid—but while some of these records have been recovered, they have proven to be out of date or inaccurate.

The Soviet and DRA forces also placed land mines on tracks through the mountains between Pakistan and Afghanistan, and many of these were remotely emplaced by helicopter and airplane. None of these minefields were marked or mapped.

The anti-personnel mines employed by the Soviet and DRA forces included but was not limited to the following:

• POMZ—a multi-directional metal fragmentation mine mounted on a short wooden post above the ground and detonated by a trip wire.

• PMN—a blast plastic cased mine usually buried below ground and detonated by a person or animal stepping onto a pressure plate.

• OZM—a multidirectional mine containing ball bearings which is typically buried and then 'jumps' to approximately 1 metre in height before detonating. It is detonated by trip wire or by a command detonation device.

- MON—a directional plastic cased mine containing ball-bearings set above ground and detonated using a command detonation device or by a trip wire.
- PFM—a small plastic cased blast mine dropped from a helicopter or airplane and detonated by a person stepping on it.

Mujahideen Mine Warfare

Mine warfare is a favourite technique with the guerrilla. Mines are a relatively inexpensive way to attack personnel and vehicles. The mujahideen mostly used anti-tank and anti-vehicular mines in the main roads and supply routes of Soviet and government troops to reduce their mobility and cut short their supplies. Most of these mines were laid in and around the provinces bordering Iran and Pakistan, and alongside the Salang highway connecting Kabul with the former Soviet Union.

When the mujahideen did employ anti-personnel mines, they preferred the directional mine (similar to the US claymore mine). The mujahideen preference for home-made mines in metal cans made it easier for Soviet mine detectors to find them. The tendency for curious troops to cluster around a newly-discovered mine is not uniquely Soviet, and the Soviets eventually trained their engineers to quit clustering around mines.

The mujahideen usually combined demolitions and mining with other forms of offensive and defensive action. They usually covered their mines with direct fire weapons. The mujahideen seldom left their mines unattended if they were located a distance from the border and a ready supply of mines. After an ambush or fight, they would often dig up their unexpended mines and take them with them to the next mission.

During the war, the Mujahideen were supplied with many types of foreign anti-tank mines. Often, the Mujahideen would stack three anti-tank mines on top of each other to guarantee a catastrophic kill. Many Afghans are inveterate tinkerers and they preferred to make their own antitank mines from unexploded ordnance and other antitank mines.

The anti-tank and anti-vehicular mines employed by the mujahideen included but was not limited to the following:

- TM series—a Soviet manufactured metal cased blast anti-vehicular mine usually buried and detonated by a vehicle driving over a pressure plate or bending a tilt rod.
- TC Series—an Italian manufactured plastic cased shaped charge anti-tank mine usually buried and detonated by a vehicle driving over a pressure plate.
- PGMDM—a UK manufactured plastic cased blast anti-vehicular mine designed to be remotely emplaced but usually buried and detonated and detonated by a vehicle driving over it.
- Mark series—a UK manufactured blast metal cased anti-vehicular mine usually buried and detonated by a vehicle driving over a pressure plate.

Afghan Civil War 1989-1992

The Afghan Civil War 1989-1992 began with the Soviet withdrawal from Afghanistan on 15 February 1989 until 27 April 1992, the day after the proclamation of the Peshawar Accords proclaiming a new interim Afghan government which was supposed to start serving on 28 April 1992.

Mujahideen groups, some of them more or less united in the Islamic Unity of Afghanistan Mujahideen, in the years 1989–1992 proclaimed as their conviction that they were battling the hostile 'puppet regime' of the Republic of Afghanistan in Kabul.

In March 1989, the mujahideen groups Hezbi Islami and Ittihad-i Islami in cooperation with the Pakistani Inter-Services Intelligence (ISI) attacked Jalalabad but they were defeated by June.

In March 1991, a mujahideen coalition quickly conquered the city of Khost. In March 1992, having lost the last remnants of Soviet support, President Mohammad Najibullah agreed to step aside and make way for a mujahideen coalition government.

One mujahideen group, Hezbi Islami, refused to confer and discuss a coalition government under the Pakistani sponsored Peshawar Peace Accords and invaded Kabul. This kicked off a civil war, starting 25 April 1992, between initially three, but within weeks five or six mujahideen groups or armies.

Afghan Civil War 1992-1996

The Afghan Civil War 1992-1996 occurred between 28 April 1992, the day that a new interim Afghan government was supposed to replace the Republic of Afghanistan of President Mohammad Najibullah, and the Taliban's conquest of Kabul establishing the Islamic Emirate of Afghanistan on 27 September 1996.

On 25 April 1992, a civil war ignited between three, later five or six, mujahideen armies, when Hezbi Islami led by Gulbuddin Hekmatyar and supported by Pakistan's Inter-Services Intelligence (ISI) refused to form a coalition government with other mujahideen groups and tried to conquer Kabul for themselves. After four months, already half a million residents of Kabul had fled the heavily bombarded city.

Over the following years, several times some of those militant groups formed coalitions, and often broke them again. By mid-1994, Kabul's original population of two million had dropped to 500,000. In 1995–96, the new Taliban militia, supported by Pakistan and ISI, grew to be the strongest force.

By late 1994, the Taliban had captured Kandahar, in 1995 they took Herat, early September 1996 Jalalabad, and eventually late September 1996 they captured Kabul. Fighting would continue the following years, often between the now dominant Taliban and other groups.

Chapter six
Working in Afghanistan

The second five contingents still continued to oversee the training courses at Risalpur but over time this responsibility was taken over by the Afghan instructors who had been trained. The UNMCTT members were increasingly used to conduct quality assurance visits to demining teams working in Afghanistan and to also run refresher courses for them in-situ. The last contingents ran a number of demining training courses in Herat in the west of Afghanistan.

> 'It was during the first month or so of my posting [February 1991] that I actually travelled inside Afghanistan and, like a kid in a lolly shop, I wanted to see and do everything, largely ignoring the dangers. One day I travelled with nine of our staff into the south-eastern part of the country to see some real demining taking place. First, we loaded up four Nissan Patrol vehicles, each equipped with VHF and HF radios, as in the 1990's satellite telephones were still huge, cumbersome affairs and very expensive to buy and use.
>
> We headed out of Peshawar and into the tribal areas after first passing a check post of Pakistani border guards that was located well to the south of the main border crossing at Torkham. After a couple of hours driving into the mountains, we came to the border check post. After some discussion between our

team and the border guards, we were let past. It was clear from the facial expressions of the border guards that they didn't see many red-haired foreigners dressed in shalwar kameez coming through their post.

At one point, just after passing the border guards, we paused on some high ground in the mountains and looked over a wide valley, with a clear track seen in the far distance. With my military eyes scanning the horizon I saw two Soviet vehicles, a T55 tank and a BRDM scout vehicle, destroyed and now stationary beside the track in the distance. I admit, even today, a smile emerges on my face and I can feel the eagerness yet that little bit of concern I had going into this wild and risky place.'

Graeme Membrey, Technical Advisor
and first Australian Army member to travel into Afghanistan

'I looked across at the two Pashtuns riding in the car with me and thought of the hundreds of deminers working out there. They were a proud and good-looking race of people. Most of the men were tall and well-muscled with an athletic presence about them. They had Aryan looks with thick, dark eyebrows and an ostensibly strong nose! All had lengthy hair which was the style of the day and wore the mandatory Islamic type of beard that was thick, relatively long and almost always jet black. Typically, they would wear some form of headdress, either a turban, Chitrali hat, or more often a turban tied around their heads.'

Graeme Membrey, Technical Advisor

'There were surprisingly very few other cars on the road as we continued to drive and I guessed we only passed three or four vehicles heading towards us, with about the same going in our direction. All were beaten up, old Toyota pick-ups, mostly filled with armed men, though we also saw a few of the very old Yak 4x4s, which were Soviet era vehicles and resembled those seen in WW2 movies. As a military guy who had studied Soviet military machinery, aircraft, ships and vehicles for several years without ever actually seeing any, the T55, BRDM and these Yak 4x4s were a real treat'

Graeme Membrey, Technical Advisor

The border crossing at Torkham in 1992.

'At about 5.30 pm we all went inside for the ubiquitous formal greeting and cups of chai. As the general banter and standard complimenting continued, some huge, somewhat scary looking men entered the room. However, to my surprise, they displayed the humblest of manners and started to lay out plate, after plate, after plate of wonderful local foods. The men trod gently across the matting on the floor, always bent down to maintain their humility, as they placed out the food and poured each of us cups of tea. They didn't look anyone in the eyes, except if they were spoken to. They appeared, whilst performing these acts of submission, to be extremely humble and accommodating. This type of behaviour, from grown and mature fighting men, never ceased to amaze me throughout the country although it was a cultural issue for them and never seemed to be demeaning in any way.

As this wonderful loading of food was laid before us, I admit I was starving as we had left Peshawar at about 10 am and I hadn't eaten anything of substance all day. There was roast chicken, meat (I foolishly though it was lamb, but as would become usual, was goat), rice, more rice, tomatoes, onions, and flat bread the size of a rugby jumper. After a quick Islamic prayer, we began to eat. As was to become the norm, I was encouraged to eat by all those around me. If I paused to catch mt breath, another leg of chicken or pile of rice was pushed in front of me. Now, please understand, I'm not so tall, but I can be a big eater and this late afternoon or early evening, I really ate my fill. I washed down the goat meat with green tea (chia subz), chomped on the chicken pieces and swallowed rice like water. The cooked potato and the salad dishes all took my fancy as did the kebab meat, the flat naan bread, the local yoghurt and the fruits. By now I was pickled and more than a little bloated. I had eaten my fill and now felt 'fat and happy.'

Graeme Membrey, Technical Advisor

A collection of 'war booty' at a Mujahideen compound near Khost in 1992.

'The minefields were surveyed and marked with rocks painted red. In the early days they were marked with steel pickets and wire but these items were very useful and valuable and soon disappeared. But nobody wanted more rocks so once they were placed and painted, they usually stayed in place. When an area was cleared, we painted the red over with white paint.'

Graeme Membrey, Technical Advisor

'The PFM-1 'butterfly' land mine is an extraordinary looking device in a bright green or sandy colour that can appear 'toy' like, though this is not its intended design or deployment. It is a small plastic device about 120mm in length and 60 mm in width, with a fusing tube of about 20 mm thick. It has a small tube through the middle with a metal rim that holds the fuse and actuator. Both sides of the PFM-1 are similar, though one is thin and of just hard plastic while the other is eight mm thick and holds a liquid form of explosive. The hard wing aids in this air-delivered landmine as it flutters down to the ground; they are deployed out of canisters from war planes and helicopters in groups of 122 mines all packed together. The explosive in the other wing is liquid so that, when stepped on, the liquid squeezes into the actuation tube and the device detonates. The results typically end in the loss of much of a foot, or hand. I am sure kids did pick these light green mines up, thinking they were toys, but again the design and deployment was not meant to lure children to them.'

Graeme Membrey, Technical Advisor

'As the Technical Advisor to ATC, I was required to do anything and everything. This included drafting and preparing the final version, including correcting all the English, of the monthly and annual reports and the regular donor submissions. It also included the clarification of those perceptions on land mine misuse. This was not always as easy as it sounds and at times was downright confrontational as some groups seemed to want to use and manipulate these perceptions in order to gain greater funding.'

Graeme Membrey, Technical Advisor

'The roads, as we travelled north, rose up into the mountain reaches of Khost and we plied through windy tracks that were often diverted due to land falls. Small groups of kids, about 7-13 years of age, pretended to be working ferociously to repair or clear parts of the road expecting to receive a few Afghanis, which was the local currency. These groups seemed to be everywhere and so, often, one of us would drop a couple of hundred Afghani notes and the kids would dive and fight each other for them.'

Graeme Membrey, Technical Advisor

Abandoned artillery piece and live rounds near Khost in 1992.

'We came across three young men walking along the track with huge rocket pods and bomb casings strapped to their backs. They told us that they collected these remnants of war and took them home to cut them up with handsaws before selling the metal. They said that every month a truck from Peshawar drove around to buy larger pieces. I warned them that cutting up explosive ordnance with a hacksaw was dangerous and they shrugged their shoulders and replied 'What else can we do?' A day or two later we heard a loud, deep explosion in the distance. Sadly, it turned out that two of the metal merchants we met had been killed cutting up an unexploded bomb.'

Graeme Membrey, Technical Advisor

'In camp one evening a group of about 20 mujahideen on horseback came galloping into the compound. I was a bit concerned that they had somehow heard about the foreigner and had come to kill me—or worse, take me prisoner. But thankfully for me their purpose was to get medical care for one of their group who had been wounded. His arms were both heavily bandaged and his head was lolling about. As our 'doctor' looked after him we learned that the wounded man was the group's commander. They had been laying landmines to block advancing government troops and one of them had detonated. While his hands had both been blown off, we managed to keep him alive and after a few days he was driven back to Peshawar. The man was very lucky to be alive, but I think of his circumstances: one minute, he was the leader of fighting men, then in a second or so the blast had turned his future. No longer was he able to feed himself, or even go to the toilet.'

Graeme Membrey, Technical Advisor

'Of particular interest to me was the flail machines that operated in another part of Khost Province. They were armoured monsters that looked something like giant, enclosed farming tractors with a rotation axle out in front of them. Attached to the axle were huge chains with steel balls welded on to each of them. As the flail would move slowly forward, the axle with the chains would spin at high speed. The chains and their heavy steel balls were flung around and around, pulverising the ground in front of them. Of course, this rapid pounding would detonate or destroy any landmines that might be laying there. The armouring of the tractor was designed to protect not only the driver, but the engine and other running gear of the enormous machines. The flail machines in Afghanistan were designed and developed especially for civilian, humanitarian demining purposes. The flails used by ATC were called Aardvarks and this was an apt name as they really did look like such animals when in operation. But the machines were subject to a lot of operational downtime due to the very demanding conditions they worked in.'

Graeme Membrey, Technical Advisor

A mine flail in operation near Khost in 1992.

'As we were driving along a track in Afghanistan, we came across four armed Mujahideen who ushered us to the side of the road. They seemed anxious and after a chat between the Afghans we agreed to pull our vehicles fully off the road—which I knew to be inherently dangerous. I wasn't sure what was going on and then it suddenly became clear. A mighty roar was heard and a huge black belch of smoke rose from over a small but steep hill about 30 metres ahead. To my amazement, up and over the

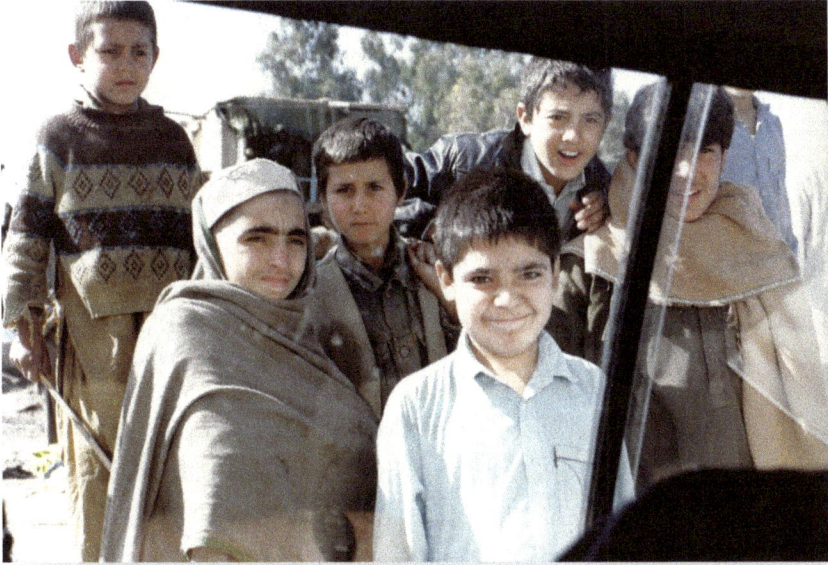

Afghan children at Khost in 1992.

A wrecked Soviet T-34 tank near Khost in 1992.

crest came an old but fully operating T-55 tank. The designation
T-55 indicates that the original tank was designed in 1955. New
models of the same design were of course made, but none had
been so since about 1975...so this was one old tank. The T-55
was carrying about five or six more mujahideen on its exterior
and it belched and burped its way over the crest along the track
towards us. I was in awe. By now I had seen many destroyed and
inoperable tanks, but here in front of me was a combat-ready
(if not old and rusted) tank from the Soviet times returning
from what they told us was a successful operation. What that
meant and what had really happened, I never knew, but cries of
Allah Akbar (God is great) were yelled and hollered by all the
mujahideen and similarly responded by all our deminers.'

Graeme Membrey, Technical Advisor

'I had a small hand-held video camera and would film my
surrounds every so often. I was filming the evening prayers at a
demining site camp one evening and panning the camera around
when I bumped into a solid object. I pulled down the camera
to see what I had hit to find a tall Afghan standing close by and
watching me with great interest. After my initial shock I greeted
him but he seemed only interested in my camera. So, I gestured
to him to come and see what I had been filming. I turned the
camera around, opened up the small viewer screen and pressed
the 'play' button. To my amazement he looked at the monitor and
then his head thrust back and he quickly stared back at me with a
guffawed look on his face. Clearly, he had never seen an operating
video camera before. He again looked at the recording in the
small viewer panel and then lifted his eyes to see what had been
shown on the screen. But of course, the person walking past in the
video was now nowhere to be seen. The young beared Afghan was
perplexed, and it was fun to watch his amazement.

Seeing his reactions, four or five other Afghans came over to see
what was going on. I gave them all a chance to first look at the
scenery in front of us and then at the recordings on the display
screen. All of us laughed and laughed, not at anyone in particular
but with each other, as none of them could really understand
what they were seeing. Very soon though, they got the hang of

what was going on and after just a few minutes of this, I allowed them to film with the camera. Interestingly, they were all very careful in handling the camera and no doubt saw it as a privilege. Each would film another then they would join and watch the 30 seconds of footage, laugh hysterically and pass it on to the next person. Probably, after 15 of twenty minutes of this, they all elatedly shook my hand and walked off, arm in arm, still giggling and enjoying the event.'

Graeme Membrey, Technical Advisor

'One night in Afghanistan my sleep was disturbed by the thumping sounds of heavy artillery shelling as it exploded a few kilometres away. I got up and saw it was 2 am. I walked to the door leading into the compound and was surprised to see the guards still asleep and nobody in sight. So, with some hesitation, I walked a little bit into the compound as the no-so-distant thunder of artillery fire continued. After a few minutes I slunk back to bed. Later in the morning dawn, at breakfast, I asked if anyone else had heard the gun fire and the general response was, 'Yes, it happens all the time.' Nobody seemed to think this was abnormal so I put it down as just another lesson learned in story about a foreigner in Afghanistan.'

Graeme Membrey, Technical Advisor

'We were driving along a narrow track in Kunar Province when a truck sounded its horn behind us. We pulled over and as it passed, I looked up from the passenger window to four men in the front cabin all with typical long beards. They recognised me as a Westerner and immediately the truck pulled up in front of us. We stopped suddenly and in the back of the truck now looking down at us were at least twenty Salafist fighters. Salafists are members of an ultra-conservative movement of Sunni Islam and they were well known for their zeal for jihad. Three Salafists from the cabin jumped down and ran towards our car, rifles in hand. Several more then leapt down from the back of the truck and soon they were at my door with teeth barred and angry looks in their eyes. They tried to grab the door handle to wrench it open, but I had instinctively locked it as they came towards me. My window was slightly down and I could hear them yelling and

*Barry Veltmeyer and Dean Beaumont enjoying dinner
somewhere in Afghanistan in 1992.*

cursing as they tried to get me out of the vehicle. AK-47 barrels
were levelled at my head through the gap in the window and the
aggression was palpable. One of my ATC associates got out of
the vehicle and put himself between my door and the Salafists
but they pushed and shoved him away. Just as things started to
become uncontrollable, and the fear in me was rising, one of the
other ATC fellows slowly got out of the car and moved into the
frenzied mob. All the while, more armed Salafists kept jumping
out of the truck and joining in the melee. The ATC fellow was
a big hard-looking Afghan and although he had sat silently in
the back seat of the car until now, he suddenly raised his arms
and shouted at the Salafists. Just like in the movies, an instant
response came about and all went immediately quiet. This guy
was clearly and former Mujahideen fighter and knew how to
handle a group of angry men. As all eyes flashed around towards
him and turned up to meet his natural height, he identified
himself and then spoke politely to the large grouping. As he
did, I could see he identified the mob leader and began talking
directly, and only to him. The Salafists were still agitated though

Animal bones besides the road near Gardez in 1992.
An indicator of potential mines.

now deadly quiet, but they appeared to want to hear quickly what tour deminer was saying and then get back to the task of pulling me out of the vehicle. As he spoke, some of the general talking started again and then some yelling and gesticulating. I was watching with extreme interest to all that was happening on the other side of my thin window—as you might expect. I was feeling more than a little vulnerable. But again, the deminer raised his hands and spoke directly to the Salafist leader. Then the leader raised his right arm and the talking stopped. A kind of uneasy calm descended on the mob. More talk and pointing, mainly towards me, continued, but now only between the large deminer and the Salafist leader. This back and forth continued for several more minutes. After what seemed like an eternity, but may have just been ten minutes or less, the deminer seemed to have made his point and the Salafist leader appeared to tell his underlings to return to the truck. Unwillingly, or so it appeared, each man drew down his rifle and skulked back to the truck. The Salafist leader and the deminer continued to speak with each other for a few minutes before they both shook hands and embraced in the local way. The Salafist leader never looked at me and just turned and walked back to the truck cabin. Each of the Salafists in the back of the truck maintained a steely glare at me as the truck slowly moved off. When the large deminer got back into our vehicle I reached back and shook his hand firmly and thanked him sincerely. He just looked at me sternly and nodded. Later, I made sure he was rewarded for no doubt saving my life.'

Graeme Membrey, Technical Advisor

'We arrived at a demining site where there had been a mine incident earlier that morning. As I was getting a brief on what had happened on of the team walked up with a something wrapped on an old, soiled towel. It was presented to me and as I took it, it felt like a flat football. The deminer slowly unwrapped the towel to reveal the severed foot of the deminer that had been injured earlier in the day. I gasped and almost dropped the foot to the ground but held on as a mark of respect and personal resolve. The bones of the lower leg had been severed by the blast and the rest of the foot was just a flattened mass of meat with pulverised bone parts sticking out of burned skin. It was a disgusting thing to review but I was intrigued to see what effect the anti-personnel mines had on human bodies. As was custom, the foot was buried shortly afterwards. The injured deminer was given first aid and driven back to hospital in Pakistan. He survived the ordeal and I was pleased to meet him some months later.'

Graeme Membrey, Technical Advisor

'Our driver had his window down and must have heard something. He slammed on the brakes and pulled off the road to park under a tree. At first, I had no idea what he was doing. We all bounced around the vehicle and of course, going off the road risked hitting a mine. Then he leaned forward and peered out of the windscreen. Two small dots sped across the horizon. Jets. Regime jets. Looking for Mujahideen vehicles. We had big blue UN letters on our vehicle but it was white and would be easily mistaken. The jets would must have been old worn but they were still highly dangerous.

I decided that that if they did see us, it would be better to be away from the vehicles than in them. So, I got out and jogged 50 metres away and settled down under a couple of trees in amongst some large boulders. Then I saw everyone else doing the same thing, though half of them ran to be with me. They later told me that they followed me because they were sure that I knew what I was doing. I wasn't so sure at the time, but was glad to have their company in our makeshift hideaway.'

Graeme Membrey, Technical Advisor

'My second last trip into Afghanistan in late 1991 was to visit the demining teams working in Badakhshan Province. This is the very northernmost province of Afghanistan way up in the Hindu Kush mountain range. The Province includes the 'Wakhan Corridor' which extends east to the Chinese border. Legend has it that this corridor of land was agreed so that the British and Russian empires would not have to share a border. Badakhshan Province was the stronghold of the Northern Alliance during the Soviet-Afghan war. It is populated mostly by Tajik and Uzbek peoples. Getting to the Province itself was an adventure in itself driving along one-lane dirt roads through passes between the increasingly steep mountains. After spluttering across the highest pass at 5,800 metres we dropped down towards large flat grassy meadows peppered with wonderful deep blue coloured lakes. During winter it is impossible to drive into the Province and the population essentially hibernates. I ended up spending almost three fascinating and memorable weeks with our resourceful and stoic deminers in this secluded corner of the world. They had an enormous task to clean up tons of aerially delivered mines, rockets and bombs—which were still being periodically delivered by regime aircraft. I also got the opportunity to brush up on my horse-riding skills as this is the main way of getting about.'

Graeme Membrey, Technical Advisor

'Like military drills, we taught deminers to lie down when they were using a prodder to locate land mines. This way, if the mine was accidentally detonated the blast would be directed up and out—so the deminer would be less likely to be injured; even though he would still get burns to his face. However, lying on the ground was not 'natural' to the Afghans and also risked the deminers getting drowsy in the sun. So, we found that many deminers squatted as they prodded—a much more natural position that they could remain in for hours at a time. This position, however, would put them in the direct path of a mine blast. It took a long time for the message about the risk of squatting to be understood across the organisation but eventually we had them all lying down when prodding. As an added safety measure, we also procured ballistic face shields and helmets that would protect their faces and heads in the event of an accidental blast. Slowly but surely we were able to reduce the risks that our deminers were exposed to.'

Graeme Membrey, Technical Advisor

'No demining work was performed on Friday—the Muslim holy day. Instead the deminers washed their clothes, did other

Mujahid posing with an effigy of a Soviet soldier near Khost in 1992.

Damaged DRA aircraft near Khost in 1992.

basic chores and walked to the few remaining shops to buy any specialties they could find. At about 11 am, I was sitting and reading my book in a section of the bombed-out school house where we were camped. A massive roar ripped through the air. I leapt up and saw the rest of the men near me do the same. We all instinctively moved to areas where we were protected from the direction of the noise—hiding our bodies against the walls or behind piles of rubble. As we looked out, one of the deminers pointed to the low sky and we saw a fighter plane undertaking a tight turn. It was probably several kilometres away but we were still wary. The low sun was shining towards us and we had difficulty keeping an eye on the plane. Then another deminer yelled and pointed towards a second fighter also banking in a similar line to the first. The noise, as the aircraft banked away, was incredible, but it quickly died down as the planes turned and headed back towards the city again. Clearly, they were on a bombing run against some target here in the city. As we watched, the planes zoomed down to a low altitude of perhaps 300 metres high, then suddenly there was an enormous detonation of dust and dirt leapt into the air, higher than the altitude of the plane. We heard the deafening noise as the sound waves of the detonation reached us. When the dust began to subside, we could just make out the second jet coming in before another enormous explosion. I admit, the several times I have witnessed aerial bombardment or been subjected to it, it always freezes my nerves. The problem is that you can hear and often see the aircraft, but you really have no idea when or where the bombs are going to explode. It is an eerie feeling I don't think I will ever get used to.'

Graeme Membrey, Technical Advisor

'Chris Coles and I accompanied by Graeme 'Chooka' Membrey from ATC were the first UNMCTT members to officially cross over into Afghanistan. We drove from Peshawar to Parachinar and crossed over into Paktia Province and worked from the ATC post in the Chamkani District. There were two teams operating out of the area working in a large minefield. It was great to see the teams at work and quite

Captured Soviet tanks near Khost in 1992.

sobering when a mine or UXO was located. This was where the rubber hit the road and we can all be rightfully proud of the roles we played in training these teams to get them to that level of operational readiness. Whilst some of the deminers had previous military service most were civilians. Sadly, there were deaths resulting from mine incidents, but these were not in vain and whilst it is hard to qualify and quantify the difference it made to many thousands of lives there is no doubt that it did.'

Ben White, 5th Contingent

'Joe Cochbain, Adrian LaFontaine and myself had gone into Kunar Province and been driven to Assadabad where a 'Scud' had allegedly struck a building full of ammunition and exploded. Highly likely it was someone 'durrying-up' or even worse in amongst the ammo, but there was ammunition of all types and origins strewn around the place. There were several 'Mujahids' who saw us poking around and were not happy. Our interpreter Engineer Hafiz said that they were Arabs and that we should leave as they were not fond of us or us being present. We did.'

Ben White, 5th Contingent

'Many of us would find out that there is a difference between diarrhea and dysentery. I remember standing alongside a Team Commander in the middle of a minefield as he was preparing a counter-charge to deal with a mine when my stomach suddenly erupted into a 'gaseous' mass of pressure that told my bowels that they were opening regardless of whether they were ready or not. I get the Commanders attention and in a combination of broken Pashtu and English. I determine the safest and shortest direction to an appropriate safe area. I get there an manage to get the cord undone on my local garb just in time to release what was nothing but a mass of stinking liquid. I squatted there with a feeling of relief for a few seconds and then look around to see two old Afghan men who were tending to their crop about 10 metres away staring at me in total shock and disbelief. Embarrassment 1, dignity nil. The difference between diarrhea and dysentery is related to time and viscosity. I am sure that there are many of us with similar stories and that was one of a few incidents that I would personally have.'

Ben White, 5th Contingent

Marcus Fielding and Mujahideen on a Soviet tank near Khost in 1992.

'In our time the Afghan government jets were still active in Afghanistan. You could hear them flying around every so often and then you would hear a bomb going off in the distance. Sometimes if they got too near, we would pull our vehicles up underneath a tree so as to not attract their attention and get accidentally bombed—although the pilots probably couldn't see or read 'UN' at that height and speed.'

Glenn Stockton, 6th Contingent

'I deployed with the 7th Contingent and then went over to Kabul in early 1992 to start setting up a new Demining Office. We started working with a UK based charity called HALO Trust. This was the organisation that Princess Diana later visited and advocated for. They began working in areas still controlled by the Afghan Government.'

Mark Willetts, 7th Contingent

'Mick Lavers and I thought we would visit a local village near one of the demining base camps. As we arrived all these kids ran out to greet us and one lets rip with an AK shooting into the air. His muzzle was waving all over the place and I thought this will be a stupid way to die. Thankfully, we weren't accidentally shot. Later the village produced a goat for dinner and slit its throat in front of us. We had to adopt false smiles as it lay there bleating and bleeding to death.'

Warren Young, 7th Contingent

'My first trip into Afghanistan was in late November 1991. I accompanied Faisal the Director of the Organisation for Mine Awareness (OMA). It was about a 10-day road trip from Peshawar to Helmand via Quetta. We stayed overnight in the Demining Quetta house and met the team that was led by a Kiwi Captain from memory.

The trip to Helmand was through tribal country which I understand was a Taliban heartland, but they were on our side in those days. I recall a lunchtime meeting in the black rocked Baluchistan Mountains on the way to the desert where I was offered (but declined) to have a go at one of their 100mm recoilless anti-tank guns they had liberated from the

A child bashes scrap metal (including live rounds) into old tank rounds near the Pakistan-Afghan border in 1992. Note he is using a unfuzed POM-Z mine.

Russians. The meeting was all conducted in Farsi with little or no translation for my benefit. I dare say it was to get approval to cross their land to get to Helmand, or just to show off their trophy. Either way we were able to proceed and they got to show off their rifle to a foreign soldier. I was also shown a 12.5mm anti-aircraft gun at another village, there was no offer to fire this one as they were short on ammo and the regime occasionally bombed them.

To get to Helmand took another three days (two nights). By this stage I had contracted a dodgy belly that turned into dysentery by the time I got back to Peshawar. I clearly remember urgently pardoning myself in the middle of meetings or meals, then having to tiptoe through the minefields of human droppings to find a secluded location to squat, squirt and manage my voluminous shalwar kameez [sic]. I also had to seek clarity on where the real mine cleared zones were in each village before I could loosen my load. I got a bit tired of having to explain my lack of appetite to all our generous hosts who insisted that to beat the squirts one had to eat plenty of yoghurt and sour goats' milk.

On my return to Peshawar my dysentery had intensified to a state where I couldn't eat anything and was passing blood. The only place to get treatment was at the Red Cross located in the Refugee camp. I was told to bring a sample and a plastic take away container (with lid) was secured for my benefit. I thought I was near death, but it wasn't till I walked through the camp on the way to the MSF clinic that I realised how lucky I was by comparison and that I still had a fair way to go before I met my maker. The MSF doctor who was literally up to his ears in amputations, cholera, TB and typhoid cases took a cursory look at my take away container, pronounced it as dysentery and told me to take it easy and eat bananas and yoghurt to replace my gut culture and potassium. I did get better over the next week and felt guilty for wasting his time, but his tip has served me well over the years on my trips to SE Asia. I donate to MSF to this day as they do a wonderful job all around the world in conflict and natural disaster zones.

The rest of my first trip involved travelling through oceans of desert sands to get to Helmand. Our driver was the nephew of Faisal a young man probably less than 20 years of age. I got on well with him and when he first picked me up at our house he asked if I had any 'Michael Jackson' tapes. Unfortunately, I didn't, so I was subjected to 10 days of continuous Indian music that consisted of lots of sitar and wailing. Apparently, it was really raunchy stuff and they were careful not to play it in the Talib villages. The messages in the music was lost on me, I now know how effective music can be as an interrogation tool. After a week of that stuff I would have done anything to get my hands on a Michael Jackson tape. Whenever I hear sitar music, I get flashbacks of riding in that Land Cruiser in the desert. One skill I did develop from that road trip was the ability to completely tune out to offensive noises. I have no issues working in today's noisy open plan offices. My partner reckons I use it on her too sometimes! I have no idea what she is talking about.

One morning the young driver was showing off to me as we were driving along and got us bogged in a sand dune. We all got out to assess the situation. All of a sudden there was a commotion. Faisal had taken the AK-47 off our Talib guard, cocked it, pushed his nephew to the ground and put the barrel to the back of his head! The young guy was crying and pissed himself. Shortly the others in our group persuaded Faisal to hand over the weapon and let the young bloke up. I was a few yards away watching it all unfold and felt powerless to do anything. I happened so quickly I was frozen and probably in a bit of shock myself. That kind of thing never happened at 17 Construction Squadron. Faisal explained later that he was angry that his nephew had risked damaging the UN donated car, wasted time and oh yeah, strayed off the mine cleared path to show off. His actions had the desired effect as we never got bogged again.

We got to Helmand and met with a UN Official from Sierra Leone who ensured us we were safe in Helmand as it was tightly controlled by the Najibullah regime and that no harm would come to us as we had paid our fees. Our Talib guide remained pensive for the next couple of days. It was interesting to see regime soldiers in their green uniforms and kepi style caps on every street corner and armoured personnel carriers manned and armed at every checkpoint.

We left Helmand after getting directions to the last Cuchie [sic] camp. They are desert nomads who often encounter mines in their travels. After getting past their security dogs, we were welcomed into the camp. Lessons on mine awareness were given to the men and children (both sexes). The women were segregated and we only caught glimpses of them. I understand that later OMA introduced female instructors to pass on the mine awareness message to women. Calico cloth mine information sheets were handed out to all that attended. I had mine made into shirts at a bazaar in Peshawar that our house keeper Mohammed arranged for me. I recently donated the shirts and full OMA packs to the Australian War Memorial. The trip and lessons were filmed by Abdul Wali, he was a nice guy who contacted me after my time at UNMCTT by writing

Guy Dugdale resting at Helmand in 1991.

to the Australian Army. He wanted me to sponsor him to Australia. I never replied, I had a young family and a career to take care of. I have no idea of his fate or the others post 9/11.'

Guy Dugdale, 7th Contingent

'Once we got the opportunity to visit the flails in operation. They were part of the program but essentially working open rural areas some distance from ATC's sites. The flail was a V-Shaped tractor that spun a drum in front of it. The drum was fitted with dozens of chains and metal cubes that either detonated or broke up the mines. It did a good job tilling the ground so was OK on rural land but could not be used in confined or undulating spaces—or on roads which it would just tear up. When the flails started up, they would generate a huge ball of dust. The driver wasn't able to see where he was going and he would need to be guided by someone observing from the side—and at a distance in the event there was an explosion. Unfortunately, the machines were not that mechanically reliable and there was a lot of down time waiting for parts and repairing them.'

Guy Dugdale, 7th Contingent

'I had to go over to Herat and then travel north up to the border to take charge of two mine clearance vehicles that had been gifted to the demining program. There was the small matter of removing the machine guns from the vehicles but that wasn't too difficult. I had a couple of days to spare in Herat and enjoyed exploring this ancient city. There were lots of very old buildings, mosques, and minarets and a magnificent underground cistern. I also found a shop keeper who had lived in Australia and was able to have a chat in Australian English.'

Mark Willetts, 7th Contingent

'Making arrangements for members of the UNMCTT to visit demining sites in Afghanistan was not a simple process. The first step was to coordinate with ATC (Afghan Technical Consultants) to determine which teams were due to be checked and of them which were working in areas that we sufficiently safe. ATC maintained good relations with the mujahideen so they were 'in the know'—not only of what was happening but more importantly what was going to happen. Once the visit was agreed ATC would then provide 'formal' advice to the relevant mujahideen group. Then going back through the UN office in Islamabad the Afghan Government in Kabul was also informed of the planned UN visit. Lastly, permits to travel through the tribal areas to the Pakistan-Afghanistan border had to be secured. Later, there was also a requirement for an Afghan

Marcus Fielding planning the demining operations at Peshawar in 1992.

Government 'visa' (a scrap of paper with a stamp on it) to be obtained. Because any of these steps could fail or be cancelled at the last minute, we planned for more missions that we could perform on the assumption that some would not go ahead. This process worked well and enabled us to keep up a reasonable tempo of work.'

Marcus Fielding, 8th Contingent

'Travelling through the tribal areas was an adventure. Getting the permit in advance always took a week of so as the old British Empire bureaucracy was still alive and well. But you could manage to get a permit on the spot if you knew the right people and had the right 'processing fee' in cash. The office in Peshawar where we got these permits was like a police station of sorts. Once through the guarded gates you first walked past prisoners who were chained up against the walls, then you went past the shipping container that held all the 'shish men' (hashish addicts) who were being 'detoxed' for a few days. Then into the duty officer's office and after a short exchange the requisite rubber stamp was officiously banged down onto your permit. On your way out you collected a couple of lads with .303. rifles or AK-47s who were then your escorts to ensure your safety. They generally didn't chat much and simply expected a 'tip' when you departed the tribal area—either coming back out or continuing on into Afghanistan.'

Marcus Fielding, 8th Contingent

'On my first trip to inspect the demining in Afghanistan we stayed with a team near Assadabad. They turned out in formation to greet our arrival and that evening we were treated to a 'banquet'. A goat had been purchased and as the sun set, we watched the team kill, bleed, dismember and cook our dinner. A few short minutes later the beast had been cooked—but evidently not well done. Not too long after retiring that evening my tummy started to gurgle. Most of the rest of the night was spent crouched over the hole in the ground in the corner of the compound that served as the team toilet. Suffice that when we headed off to work the next morning, I wasn't feeling too flash.'

Marcus Fielding, 8th Contingent

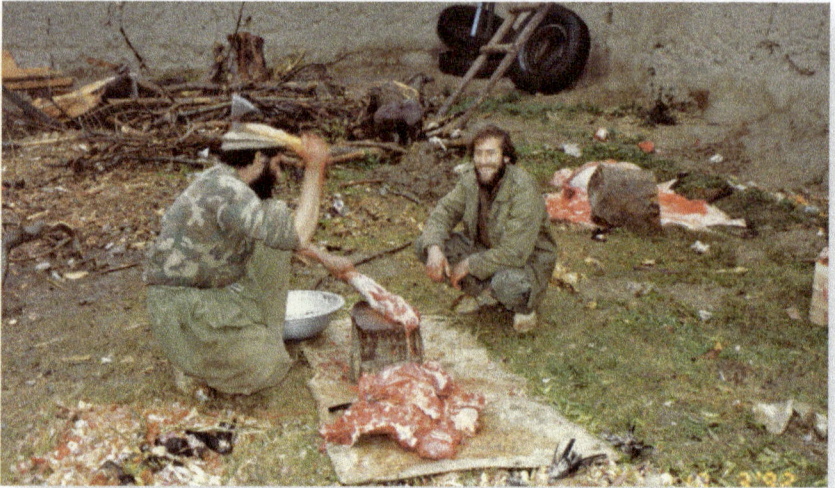

Preparing dinner near Assadabad in 1992.

An ATC Demining Team near Khost in 1992.

'I recall visiting an ATC demining team in the field with Paddy Johnson—the ATC Technical Advisor. He was highly regarded and as we approached their base camp, we could see the 30-man team all lined up in three ranks. As soon as the car stopped the commander called the team to attention and Paddy jumped out. He returned a wonderful salute from the commander and then inspected the troops. Once that was complete, he addressed the deminers and via a translator passed on the well wishes from the ATC boss and relayed how well they were performing. It was like

Marcus Fielding checking out a ZSU 23-4 anti-aircraft gun near Khost in 1992.

being back on a parade ground in an Australian Army barracks! But they all seemed to revel in the regimentation and the sense of professional ism that had been instilled in them.'

Marcus Fielding, 8th Contingent

'While we were inspecting some demining operations around Khost we got to chatting with a local family near their home. After breaking the ice, they showed us some of the weapons that had—ostensibly to defend their home and farmland. Lee Enfield .303—check. AK-47—check. A Russian helmet which one of the children put on and allowed us to take a few photos. Then the older man in the family brought out an object wrapped in a dirty oily cloth. As he peeled the folds back, we looked on in anticipation of being shown something special. When he revealed a Webley pistol, we were astonished. It was a six-round revolver. Small and with its zinc finish very worn in many areas. But it was still functioning and along the barrel was engraved 'Army and Navy Store'. Evidently it had been purchased through mail order—probably by a young British officer. How that owner had lost the revolver and where it had been for the last 80 or so years was the subject of some curiosity but we didn't seem to get a clear or detailed answer.'

Marcus Fielding, 8th Contingent

Mujahideen on a BMP near Jalalabad advancing towards Kabul in April 1992.

'Driving along a road west of Jalalabad in late April 1992 we noticed a large cloud of dust coming towards us. We pulled over to allow a small convoy of armoured vehicles head west towards Kabul. The vehicles were all former Russian Army equipment now being used by the Mujahideen. Riding is better than walking and every vehicle had a full load of Mujahideen soldiers hanging on to the vehicle for dear life. As they passed by, we exchanged waves and yells. They we cleared on their way to a new mission. A few days later—on Anzac Day—we learned that Kabul had fallen to the Mujahideen.'

Marcus Fielding, 8th Contingent

'We spent a lot of time on the road and the deal was that we would alternately listen to western music and 'Hindi' music cassette tapes so as to cater for the westerners and Pakistani/Afghan passengers. Sadly, though we only had a few western cassettes and one of them was Boney M. After hearing Ra Ra Rasputin for the tenth time I cracked it. I ejected the cassette and threw it out the window into a minefield. No more Boney M!'

Marcus Fielding, 8th Contingent

Live anti-tank rockets inside an abandoned Soviet BMP-2 armoured vehicle near Khost in 1992.

'The equipment and arms of war littered Afghanistan and was often stockpiled under the control of the local commander—thereby burnishing his credentials. They were entirely comfortable with us indulging our professional curiosity to 'have a dekko'. Tanks, armoured personnel carriers, helicopters, jets, transport aircraft, artillery pieces and rocket launchers were all there in various states of repair. And literally mountains of munitions. Small arms—rifles and pistols—machine guns and rocket launchers were typically kept in storerooms. We would occasionally take an interesting souvenir but most of the items weren't able to be brought back into Australia. Occasionally we would come across a vehicle that had been immobilised and therefore wasn't going to be put into a stockpile. These were then fair game and would be quickly stripped down—and eventually cut up with an oxy-acetylene torch for the metal to be sold to recyclers. It was probably only a few years before all the blown-up vehicles across Afghanistan ended up piece by piece back in the metal smelters.'

Marcus Fielding, 8th Contingent

*Chatting and posing with a local family near Khost in 1992.
Julian Gregson is second from left.*

'After the Gulf War the UK Army downsized and a couple of fellows joined the HALO Trust which was a demining charity working in Afghanistan. I recall hosting Julian Gregson who was a former infantry Major. We spent a few days together in Pakistan and Afghanistan as I introduced him to our way of doing mine clearance operations. He was an affable and capable chap—clearly excited about taking on a new role. Later he was assigned to their office in Kabul. The HALO Trust approach was more hands on than ours and the expatriates were physically involved in doing the demining work. Several weeks later we heard that he and another former UK Army officer were crewing a Russian tank fitted with mine clearance rollers. Sadly, they triggered a mine that must have been 'daisy-chained' to another charge and the fuel drum on the rear of the vehicle exploded. Both Greg and his associate were standing in the turret of the vehicle and were severely burned. The other fellow died at the site and Julian died some days later.'

Marcus Fielding, 8th Contingent

'We went all the way across to Gardez and Ghazni on one monitoring mission. Both towns had huge mud-walled fortresses in the middle of them. The fort in Ghazni was famous for being

*Marcus Fielding in Gardez in 1992. The famous Bala Hissar fort
in the background.*

a place where the British sappers had blown up the gates and
stormed the fort in 1839 during the First Anglo-Afghan War.
I recall Lee Uebergang was on this mission and he got crook as
a dog in Ghazni. We decided to drive home early and Lee was
basically out for the count the whole way.'

Marcus Fielding, 8th Contingent

'All the Afghan deminers were covered with life insurance.
This provided enormous peace of mind for the deminers as
they knew that if something went wrong their family would
be provided for. The challenge was that the insurance was
provided by Pakistani insurance companies and in the event
of a mine clearance accident in Afghanistan they were not
in a position to send in an 'assessor'. So, the arrangement was
that we (the UNMCTT members) would provide a report.
We were doing these investigations in any instance in order
to determine the cause and improve our drills and training,
but the reports for the insurance company had to be pretty
comprehensive and formal. I remember doing several of
these reports over the tour. We tried to do them on the back
of existing missions and trips but sometimes we would have
to mount a mission just for the purposes of completing an

An antique cannon near the airport at Herat in 1992.

investigation. Visiting wounded deminers was a heartbreaking duty. In order to prevent any fraud, photos of the injury or the body were required. Because dead Muslims must be buried within 24 hours the deminers team were directed to take photographs of the deceased. On a few occasions a UNMCTT member was able to get to the body before he was buried. The knowledge that it would help the family seek compensation made the difficult process of taking photos a little easier. We also had a policy that once these photos were included in the report that no copies were to be retained.'

Marcus Fielding, 8th Contingent

'Getting used to the food at the demining camps was a challenge. Sitting cross legged and always eating with your right hand from communal plates. There was lots of fat (ghee I think) and many dishes that you could only guess what they were. The demining teams bought food locally but we were never allowed to chip in to pay for food. We made a habit of bringing big bags of rice and fresh vegetables whenever we went out to a demining camp.'

Darrell Crichton, 8th Contingent

'In June 1992 Barry Veltmeyer and I went over to run our first demining course in Herat. It was a remarkable ancient city that had clearly seen its fair share of the fighting but the big

Winding up our last mission in Herat and getting ready to come home.
Darrell Crichton and Marcus Fielding at Herat in 1992.

minarets were still standing tall. We were planning on training
100 students but only got 54 in the end. Unlike over in the east
of Afghanistan there was a lot more reluctance for men to do
demining work even though the hospitals and streets were full
of kids and adults missing limbs.'

Darrell Crichton, 8th Contingent

'One particularly difficult demining problem was clearing the
irrigation channels. Afghanistan is clearly a dry country and
the locals have long since worked out how to channel the snow
melt to irrigate large tracts of land. These irrigation channels,
are essentially trenches and offered terrific cover for soldiers
to move around. Not surprisingly then the Russians and the
Afghan government troops put land mines into them to stop the
Mujahideen from using them. The trouble is that the channels
also silt up and have to be cleared by hand every year or so. So,
if the channels are to keep working, they have to be cleared
and a shovel and a foot don't mix well with anti-personnel land
mines—particularly when kids get the job. And for us to clear
them there was often years of silt built up covering the mines.
But to get the land working again there was little option than
to get in there and do our best. The worst thing was that even

after the channel had been demined, the water from a snow melt would move mines, which were plastic and buoyant to a degree, down from the higher parts of the system and the channels would be made dangerous once again.'

Darrell Crichton, 8th Contingent

'In the third week of April Danny Shaw and I were doing a training course for HALO Trust guys near Kabul. Things were starting to get a bit interesting and there was clearly something shifting with the amount of military activity going on. One day we heard that the mujahideen had captured the Bagram air base and that Najibullah [the Afghan President] had resigned. This didn't bode well and we were ordered to return to Pakistan but the only flight we could get onto was a commercial flight to New Delhi—which we grabbed!'

Dean Beaumont, 8th Contingent

'I began my time with SWAAD down in Quetta in about July 1992. One of my tasks was to relocate the SWAAD Headquarters from Quetta (Pakistan) to Kandahar (Afghanistan), which meant many road trips between the two

Mark O'Shannessy with a group of Afghan deminers near Khost in 1992.

cities along long desert roads that passed through many tribal areas controlled by various warlords. Each area was guarded by checkpoints which required our convoys to stop and pay some sort of 'homage' (bribe) to the local Shura. On one journey, in about September of 1992, our five-vehicle convoy was stopped for a routine check. The situation rapidly deteriorated when the homage price was beyond what my convoy commander was permitted to remit. The response was aggravated and swift. Before I knew why I could see commotion ahead, I was yanked from the back seat of the Landcruiser I was travelling in, and several Kalashnikovs were shoved in my back and under my ribs. A couple of henchmen grabbed me under each arm and my feet were not on the ground. I struggled in attempted resistance which gave me footing as I was marched/dragged to a darkened mud hut for who-knew-what was next to come. I was pushed and stumbled into a hut and blinded by the instant transition from the bright light outside to the darkness inside. Eventually my sight and bearings regained; I was not alone. Waiting inside was my guard. The teenage boy carried a Soviet PK LMG slung across his front with the bipod extended—he struggled with the weight and shape of the weapon on his small frame. He spoke to me and although I could not understand his words, I got his message; he wanted to search me. He unslung his weapon and placed it on the ground across his path to me and stepped over it—not a display of professional soldiering in my view. I obliged his efforts to the extent I thought I could get away with, which worked. He did not detect my prized Leatherman tool, nor my ID tags, which was pretty much all I had on me. Having completed his duty, he stepped back from his pat down, only to tread directly onto his grounded weapon, stumbled awkwardly and fell backwards onto the dirt floor. Now he was on his back, me standing over him and the weapon easily within my reach— as confused and fearful as I was, I recall feeling as though I was being held by the keystone cops and any sense of intimidation dispelled in that moment.

Although I was a little bewildered by the events, from sitting in my car to being searched by a teenage conscript, the inept conduct of my guard gave me the upper hand. I could sense his embarrassment so I let him regain his feet with as much grace as I could afford him. I think he must have been grateful for my forbearance because he couldn't get out of that room fast enough. Now I was alone and uncertain. Some time passed and I could hear further scuffling outdoors. My Turkish and Scandinavian counterparts, both wearing local dress (I was in the all too familiar UNMCTT Safari Suit on orders from Land Command) soon joined me. Tahsin, the Turk, was bleeding from the mouth but Sven, with a towering and formidable stature, was unscathed. Tahsin soon explained that the local Shura was displeased with recent UN interactions and his response was to hold up the next convoy (ours) to barter for a better outcome for his people.

In our favour, the SWAAD commander travelling with us outranked the Shura and we were released within a few hours. Apologies were exchanged and we were again on the long desert road to Kandahar; Tahsin was still a little bloody and understandably not happy.'

Mark O'Shannessy, 8th Contingent

Mark O'Shannessy and other members of SWAAD watching a lesson on mine awareness in a school in Kandahar in 1992.
Lieutenant Colonel (Retired) Tahsin Dishbudak (from Turkey) is holding the white paper.
Captain (Retired) Sven Hendrickson (from Norway) is crouched in the far right of frame.

'Among several duties I had as Technical Advisor to SWAAD, one was the surveying of prospective mine fields, registering the job and allocating SWAAD Demining Teams to the jobs. On this occasion, my driver and I had met the local Shura in a satellite village out of Kandahar. We met, had lunch (which was always a lavish affair) and discussed locations and priorities over a map. Once lunch was finished, we mounted our respective vehicles to tour some of the places we had been discussing.

We were visiting the last stop for the day. At the Shura's suggestion, as a precaution, we pulled up short on a grassy track leading to a small rise; the target area was on the other side of the slope. We parked up (herring bone style off the track) and walked the remaining few hundred meters over the crest and surveyed the fields before us. Boundaries were agreed and undertakings were made for a SWAAD Deming Team to be on site soon—importantly, our drivers were with us.

The Shura directed the drivers to run off and prepare the vehicles for our return to town—dutifully, they scampered up the hill, over the crest and disappeared down to the waiting vehicles. Within moments of their disappearing from view a loud explosion erupted from the vicinity of the cars and black smoke could be seen rising from beyond the crest—the unthinkable had happened.

The vehicle I was travelling in had been started, placed into reverse and moved no more than a foot or two. The rearwards track of the front right wheel (driver's side) had traversed an anti-tank mine. The vehicle was lying 20-30 meters from the blast crater. The driver's compartment was demolished and the driver was lying, severely dismembered, separate from the vehicle. He died on the scene despite attempts to save him.

The Shura's vehicle had not moved and was left until the area was cleared several days later. I have not been able to recall the driver's name since the incident, and not that I knew him well, he was a demining brother with a friendly outlook who looked after me the best he could despite language barriers. I was assigned to complete the incident investigation and his family received the benefit of the 'life insurance policy' that all killed or injured demining workers received for their contribution to UN demining effort in Afghanistan. The experience left me in shock for some time afterwards and one I have never forgotten.'

Mark O'Shannessy, 8th Contingent

The result of a UN vehicle running over an anti-tank mine near Quetta in 1992. The driver was sadly killed.

'For months we had worn mufti (shalwar kameez) while working in Afghanistan but since I had bright red hair and was considerably taller than your average Afghan, I figured it was a pretty poor disguise. Plus, we were driving around in vehicles with big blue letters on the side. So, I figured we should start wearing the tan uniform—which wasn't very comfortable but with a couple of badges on your arm at least it made us look a little bit more military like. Thank fully by this stage the security situation was settling down—at least where we were demining.'

Lee Uebergang, 9th Contingent

'It became fashionable for our vehicles to be taken and held for ransom. Local mujahideen commanders who wanted to get us to work in their areas figured that if they held our vehicles to ransom then we would clear the areas they wanted in order to get them back. At one point I think we had over fifteen vehicles being held prisoner around the country. It took a lot of time and effort to go and talk to these people and explain how we prioritised our efforts and that by them holding our vehicles wouldn't change that but only make it harder for us to help Afghanistan.'

Lee Uebergang, 9th Contingent

'Our Demining Team structure normally consisted of two Australians, four Afghan Instructors and two Afghan drivers. We would head across the Eastern border either through the Khyber Pass from Peshawar, or if down South, from Quetta and into Kandahar. We would always travel in a minimum convoy of two UN marked vehicles with trailers. These would be packed to the hilt with water jerrycans and emergency rations, tents and tarpaulins, training aids and demining equipment, extra fuel, and generators. Most importantly we would always carry a cash reserve for use in emergencies.'

Craig Egan, 9th Contingent

Mark O'Shannessy briefing at Quetta in 1992.

'The level of risk during our demining missions into Afghanistan would not be accepted today. Today's combat focused medical training, the quality of individual medical equipment and kits, and the application of the 'golden hour' simply did not exist at the time. We knew the risks were extremely high, our first aid and medical kits were rudimentary, our means of communications vulnerable, and initial medical evacuation reliant on the local treatment facility; we were isolated in many instances, and I know this worried our leadership. As such, when commencing a new mission our routine was to immediately visit the local medical treatment facilities, establish their capacity for triage, and establish our initial medical evacuation procedures. I can't recall a deeper and dedicated medical evacuation capability; I'm sure there must have been?'

Craig Egan, 9th Contingent

'As part of our Demining site inspections in Afghanistan we would evaluate the Demining Teams' medical evacuation plans. I recall there would normally be a UN funded ambulance always on site with qualified medical attendant and possibly a Doctor. In addition to the immediate site response to an incident, the medical evacuation plans were reliant on Medical facilities provided by the ICRC, Medecins Sans Frontieres, or the local Mujahedeen Hospital (to use the colloquial terminology of the time). These facilities were staffed with volunteers doing the best with what they had, living in arduous conditions, working in an unpredictable and violent security situation, while balancing a culturally complex environment.'

Craig Egan, 9th Contingent

'The efforts the Mine Clearance Program had gone to in educating the population was commendable. Understanding the environment and culture was critical to the successful application of the Program. Poor literacy levels were overcome through pictures, posters and booklets; silks and cloth used instead of paper to ensure durability and longevity; the convincing of key tribal, local and village leaders, of the direct benefits overcome the lack of radio and television in communicating the message; further, understanding the cultural and population dynamics allowed women and children to be affected directly. Sadly, despite all this effort, the medical facilities charged with providing the first point of treatment for the local population, were full of wounded children due to mine strike—blinded or loss of limb. As a result, children forever relegated to a life of perpetual poverty and lost opportunity in a harsh and unforgiving land.'

Craig Egan, 9th Contingent

'We were conducting a mine Incident investigation outside of Jalalabad. One of our Deminers had been seriously injured a few weeks prior. The Site Engineer was escorting me to the point of the detonation. With total trust in his understanding of the site, we set-off at pace. I followed with confidence, leaving the working minefield and its painted rocks indicating the boundary of clearance. We walked for about 10 minutes, until he started to randomly change directions and appeared to look uncertain. Sure enough, he had guided us into an uncleared area of the minefield. 'Mine Freeze'—we assessed the situation. We were approximately 20 metres beyond the last mine indicator—or certain cleared point. Slowly and ever so cautiously, we retraced our steps, occasionally balancing on a large semi exposed rock, and navigated our way out. After what felt like an eternity, at a random point, the Site Engineer became confident enough to say, 'Mr Craig, we are safe now'. After a quick debrief, reset, and a line of questioning he may still very well be recovering from, we continued with the task. Once again, we moved into an area without painted rocks...however, this time the blood trail indicated it was the correct site.'

Craig Egan, 9th Contingent

'We were fortunate to visit the same location in Jalalabad early and late in our deployment. During one of our early 9th Contingent missions the Demining Team had recently commenced clearance of a football sized paddock. Upon return late in the deployment, the same paddock was being tilled using an ox and plough, and crops were growing. It gave us a remarkable sense of job satisfaction to see such progress.'

Craig Egan, 9th Contingent

A group of Afghan instructors at Herat in 1992.

'Dave Mitchell and I had been in Herat for several days on a mission to explore opportunities for Demining Program expansion. We worked from an ex-Russian and regime military barracks on the city outskirts, and were accommodated in a UN common house centrally located near the Governor's precinct. This arrangement provided an opportunity to ensure our demining efforts were coordinated with and complimentary to other UN efforts, and aligned to the existing conduits needed to attain approval from local Afghanistan authorities. Noticeably, many parts of the city had been 'levelled' during years of conflict, with minefields located in close proximity. The significant destruction to outlying urban areas and the extent of mining, including the rubble, effectively denied any access and opportunity for rebuilding and reoccupation; our mission aimed to change this circumstance.

Herat like the rest of Afghanistan was filled with armed mujahedeen; however, it differed in that it appeared to be orderly, structured, disciplined and regimental in some ways—particularly with the blend of uniformed ex-regime members in the area—and stable unlike many cities in Afghanistan.

Notable historic and cultural aspects were the four minarets, the blue tiled mosque, and the abundance of 'blue glass' for which Herat was famous. Early indications were for an effective and productive mission built on the preceding work of the 8th Contingent, enhanced by the flow of refugees back to the city under UN and NGO resettlement programs, and evidenced through a relatively stable security environment.

This feeling of stability would change rapidly commencing with the discovery of mass graves in the surrounding hills of the old military barracks. With the local UN lead representative, we were called to the site to inspect and report the findings. Upon arrival we were escorted to several unmarked dig sites that exposed numerous partially clothed skeletal remains; all with a single bullet hole through the head. The area was marked and measured approximately 100 metres by 20 metres. I cannot recall how many bodies were uncovered but they lay where they fell, overlapping in a macabre domino effect of multiple layers, indicating a possible act of mass execution. Remarkably, despite the presence of ex-regime and mujahedeen, who had obvious history and ties to Herat, nobody had any knowledge of the executions except to say 'the remains were local people'. Soon after, additional UN higher representation flew in, thus the site discovery was officially reported and investigations commenced.

The graves were further investigated followed by an official ceremony involving the Governor Ismail Khan. People ringed the perimeter of the site and commenced to wail, pray and express their grief through passionate and intense speeches. This became rather unnerving due the obvious emotional impact on the local population, the potential for reprisal (some people present may have known of these mass graves), our obvious foreign origins and poor understanding of the language—Farsi was the dominant language spoken in Herat; we had been schooled in basic Pashtu. That evening the entire city commenced to publicly grieve which included community wailing of 'Allah Akbar' (my understanding meaning 'God is great') with the night sky erupting with gun fire and tracer. To our alarm and surprise, a significant concentration of fire came from a DShK 12.7mm Anti-Aircraft Heavy Machine Gun, positioned on the roof of the adjoining property to our UN house.

In the days that followed the security within the city gradually declined with rumours of a hostile warlord opposing Ismail Khan's control. This was confirmed soon enough with the definitive rumblings of battle increasing on the outskirts near the airport and the movements of tanks throughout the city streets as they were responding to the threat. The tempo of this battle increased dramatically over several days with continued tank movement and sustained artillery fire targeting the nearby Governor's residence. With the city under deliberate attack we were ordered to remain within the confines of the UN common house, along with the various UN and NGO staff, and prepare for evacuation. We were required to make several scheduled daily Situation Reports (SITREP) by radio. A vantage point on the roof of the UN common house provided us with reasonable observation above the high mud-brick wall that surrounded most dwellings in Afghanistan. From this observation point we could view the surrounding mountains and follow the blast and smoke trails of the battle.

The intensity increased gradually with 152mm artillery pieces conducting fire-missions. These guns had been secreted in large caves dug into the surrounding mountains. We watched as they were efficiently removed from cover, positioned for the fire mission, and with the accompanying blast and barrel vibration reverberating through the air, rapidly secreted back into the protection of the caves once again. These actions were followed quickly by the opposing force's counter-battery fire as it peppered the position and the surrounding mountain. After a brief interlude, these engagements would be followed in a similar manner using Multiple-Barrel-Rocket-Launchers. Hearing the high-pitched scream, observing the violent back-blast and following the rockets across the sky, I could only feel for the recipients of this brutal but effective weapon.

This aspect became poignant as the battle closed-in and the UN common house fell within the fragmentation zone of these weapons. Fortunately, we had earlier convinced most UN staff to relocate to the cellar; now due to the intensity of the barrages and the collateral strikes on our compound, we found all staff could be accounted for easily. Gradually we were able to steadily support our UN colleagues and influence their preparations for what could be several days underground. Our military training was paying-off as we advised on scheduled briefings and situation reports, and supported the systematic coordination of fuel, water, rations, medical supplies, first aid revision, and controlled personnel movements. To reduce our signature to outside observation, all lighting outside the cellar was extinguished, windows blacked-out, generators were run only for essential services such as food preparation and radio communications; however, the latter would fail us in the last few days as our ability to recharge the batteries became problematic.

Until this point, Dave and I had continued with our daily situation report (SITREP) to Mission Headquarters in Islamabad however these changed from a timed schedule to an opportunistic schedule due to the increased rate and proximity of shelling. These schedules required Dave and I to move from the house across exposed ground to the radio room located in a small external building. Due to the unpredictable regularity of the artillery fire, it was during these moves we were regularly caught in barrages of air-burst artillery. The air around us cracked and the ground erupted as fragmentation chewed at the ground. Fragments missed us by centimetres smashing and cutting into the surrounding vehicles and vegetation as we dashed to the communications room and its limited protection. I can still feel the intense heat in my hands from multiple fragments picked up off the floor as we propped below the window. That small wall space provided the cover needed as we made our SITREPs. These SITREPs were critical in communicating our circumstances to the outside world, for in Islamabad, our Mission leadership was working tirelessly planning and negotiating our options for evacuation.

Concurrent to these efforts, and with our communications systems becoming unreliable, the senior UN representative and

Dave (being the senior Australian Army representative) displayed skilful leadership, management and interactions at the local level, that would ultimately lead to our safe extraction and evacuation to the border of Iran. Over several days the forces loyal to Ismail Khan gradually regained immediate control of the City and focused on the routes leading out. A decisive point appeared to be the intervention of jet fighter aircraft from the city of Mazari Sharif, I believe controlled by Ahmad Shah Massoud. The first we knew of this was when jet fighters screamed over our head and dropped bombs in close proximity. I was outside at the time when the shockwave punched into my face, immediately followed by the large glass window behind me partially splintering over my back. Suffice to say, I made a rather quick dash to the cover of the cellar.

The jet fighter intervention, I believe, proved to be the turning point in the battle for Herat. As a 'show of force' it displayed unity to Ismail Khan by Massoud, thus the battle eased, and our opportunity to evacuate arose. Having refuelled all vehicles over preceding days, stocking them with supplies and baggage, all involved were briefed, and early in the morning we departed West for the border of Iran. This move was sanctioned by Ismail Khan and effectively provided us safe but unpredictable passage to the Iran border. The move which took several hours was uneventful. Upon arrival at the border all UN and NGO personnel moved quickly through the check point however, this was not the case for Dave and I. With an aggressive and firm arm thrust across my chest the Guards abruptly stopped us from entering the processing office. Dogmatically one stated 'you're a soldier' which alarmed me as we were not dressed in military clothing in the hope of blending in. I responded in acknowledgement to which the Guard continued with more aggression 'Why do you come to my country?' 'You're a soldier, you don't belong in my country!' Eventually, after the guards had affirmed their authority, they spoke to their supervisor who eventually allowed us to enter the processing office. Dave and I sat for what felt like hours while the administrators' held discussions, reviewed our passports, made endless phone calls and finally gave us a reassuring nod and smile. While we did not realize it at the time, the political implications of two foreign soldiers seeking what was in affect refuge, was significant. The slow processing, while somewhat disconcerting at the time, was completely understandable.

Late into the evening, we arrived at an Embassy in the city of Mashhad (Iran's Holy City). Dave and I received a personal note of welcome from the Australian Ambassador offering any assistance we required. We had been out of contact with Islamabad and Peshawar for several days, so to receive that message was very reassuring particularly knowing our leadership and families would be informed of our wellbeing and location. Circumstances over the preceding weeks had been a little grim and full of uncertainty to the point our Officer Commanding, Major Lee Uebergang, had the unenviable task of calling my now Wife, to inform her that Dave and I had been caught in a battle and had been missing for several days. No doubt due to a considerable diplomatic effort, we departed several days later for Islamabad on the UN aircraft Salam 01 to be reunited with our relieved fellow Contingent members.'

Craig Egan, 9th Contingent

'Enabled through our Afghan teammates, at times we were fully immersed with the local population. While we had military rations for emergencies, we brought a majority of our meat and vegetables from local open-air markets. These would be combined into a casserole and cooked in our pressure cooker ensuring meat was tenderised and a majority of bacteria eradicated in the process. The main dish would normally be complimented with naan bread, potatoes or rice, chickpeas, and local fruits which included pomegranates, watermelon or grapes depending on the season. For most meals we would sit on large rugs in a communal circle and share the food. Occasionally, of a morning, our driver would make us a delicious treat of sweet Afghan Milk Tea (Sheer Chai), one of my favourites. Unfortunately, due to our many stomach related problems, our appetite was normally very low. Once again, in trying to assist us, our Afghan colleagues would provide us with a salty yoghurt drink which tended to provide a temporary relief. I think it was called Doogh, similar to Lassi.'

Craig Egan, 9th Contingent

A mine awareness education sheet explaining what to do in the event of finding a land mine.
Must be read from right to left.

'We were working from an old school house on the outskirts of Jalalabad. As usual, our security was provided by the local provincial Governor or the power broker in that Region. These Mujahedeen were a mixture of hardened veterans and young aspiring fighters. Many appeared to be significantly affected by opium and other drugs. At one point I was alone in a building where our stores were kept. A young man carrying the stock standard AK-47 came in, stood in the only exit point and demanded 'tablets' for a sore finger. Knowing what he was after 'tablets = drugs', I offered to disinfect his scratch and apply a band aid, to which he became disproportionately agitated. I attempted to convince him he was a 'strong Mujahedeen warrior', and what I was offering was the best treatment if applied over several days. With this intent in mind, I moved to our first aid kit positioned at the opposite end of the room. I heard the rifle cock. I turned to face the young man only to see him in a squatting position in the entry, weapon aimed directly at me; he proceeded to place the safety catch to fire. My fight or flight instincts kicked in, I proceeded to walk toward him, our eyes locked. I motioned intimating if he did not point the rifle away, I would grab it and shove it up his arse, all the time moving slowly and steadily toward him. As I closed in, less than 3 metres apart, I gestured to push the barrel away, at which point he placed the safety catch to safe and ran from the room, never to be seen onsite again.'

Anonymous, 9th Contingent

'After a period of time, the nightly machine gun fire emanating adjacent to our buildings became somewhat normalised, as factional fighting continued between the Mujahedeen. Mine clearance had not become as normalised, but we were able to predict patterns based on experience and our own Royal Australian Engineer Mine Warfare doctrinal teachings. For example, the Russians had mined Power Lines to a density of 1 mine for every square metre (1:1 ratio). This reflected their Vital Asset status. Dubiously and possibly cruelly, cemeteries had been densely mined as well. While it may be justified from a tactical perspective, this breach of Conventions regards culturally significant sites must have had a dramatic effect on the local population. Endless remains of animal bones were scattered throughout, rotting wooden stakes would hold their lethal payload precariously within half a pace of the safe route, waiting for a strong breeze, an inadvertent rodent or unsuspecting person to upset the earth around it. Tripwires would be intertwined, triggered by tumbleweed as it unpredictably blew through the mined areas. These were the daily challenges our Demining Teams faced. The mine signs aided in identifying the threat, the courage of our Afghan Demining Teams ultimately reduced the hazard in the most difficult of circumstances.'

Craig Egan, 9th Contingent

Afghans with a Soviet Hind helicopter near Khost in 1992.

'On our last mission, we were tasked at late notice to escort some Australian visitors (I can't recall if they were Officials or NGO representatives) to a minefield. They accompanied us to the site which was adjacent to a mined cemetery. We moved along the safe lane, animal bones, trip wires, and POMZ mines on rotting stakes within half a pace off the track. To my alarm, one of these visitors pulled out a boomerang from his backpack and went to throw it. I intervened explaining the shear recklessness of the act; the young man persisted saying 'come on, how many people can say they have thrown a boomerang in a minefield?' He persisted foolishly only to find his boomerang replaced by the intervening determined grasp of our trusted Infantryman Sergeant Glenn Close. Glenn, in typical Infantry style, swiftly moved and exercised his unarmed combat skills on the young fellow. Suffice to say, he was temporarily distracted from his boomerang endeavour, shifting his priority to the preservation of his throwing arm. We found him most cooperative as he was unceremoniously escorted from the site.'

Craig Egan, 9th Contingent

'During a visit to our demining teams in Qalat, we were invited to the Qalat Fortress by the Provincial leader. The Fortress is atop a large hill dominating the skyline and overlooking Highway 1. Its foundations date back to the forces of Alexander the Great, with evidence of its more recent military history of Russian occupation through abandoned command wagons, motorcycles, rocket launchers, and the high walls ringed with precariously balanced POM-Z Mines interlocked with tripwires. I had read books about the Russians spending months behind these or similar walls, under constant attack, losing their minds as their resupply conveys were decimated by the Mujahedeen.'

Craig Egan, 9th Contingent

'We were in Qalat in August and there was no simple way of escaping the harsh dry heat without the benefits of modern cooling systems; well, so I thought. Stopping at mujahideen check point, we were invited for chai and watermelon by the local commander. We entered a thick-walled mud hut then proceeded through a room that led us to a lower level. This room was 75% below the surface leaving sufficient space above ground for windows. Smaller vents had been dug below the surface. Remarkably, it felt like it was air conditioned. Ingeniously, the lower vents were stuffed with spinifex and every few minutes would be splashed with a small amount of water. This created an air circulation that extracted the heat via top windows and vents, replacing it with 'water cooled' air, drawn through the bottom vents.'

Craig Egan, 9th Contingent

'Our Contingent's first deployment into Afghanistan was to Kandahar. Our mission was to write a report recommending workable clearance techniques for mined areas in creeks, drains and thickly vegetated and overgrown 'green belts'. We departed Quetta in Pakistan by road and crossed into Afghanistan through the Chaman Border Post. From this point, as previously arranged, we were accompanied by two 'Hilux' type utes full of armed mujahideen, compliments of the local provincial authority. The ute with the centre mounted machine gun went to the front, the other to our rear. These would be replaced with new teams as we passed through other mujahideen check points; in effect it was a point of handover between mujahideen areas of responsibilities. After travelling on relatively good quality roads, and occasionally bypassing a destroyed bridge, several hours later we arrived in Kandahar and were escorted immediately to the Provincial Governor's palace. We were official guests of the Governor therefore we were allocated a large room to bed down, and directed not to leave without the Governors approval. We were under constant protection by the Governor's security element. This level of protection while perplexing initially soon made sense when considering the atmospherics.

Kandahar had a repressive feel to it generated by an overtly aggressive and intimidating mujahideen presence akin to the 'wild west'. Mujahedeen in utes with mounted machine guns and blue lights constantly patrolled the streets and were most prevalent at the major roundabouts throughout the city. Every morning we would be escorted to the mined areas to conduct our reconnaissance tasks. We were briefed to stay on the same foot track and never to deviate. This was cleared and deemed the safe route as evidenced by the demining teams' use over several months. Ironically on our last morning and reinforcing nowhere is entirely safe, we heard the blast; on this very track a deminer stood on a land mine as they headed to work. Considering the overt hostility, levels of security, and the fact these lanes and tracks had been cleared and used for several months, I could not help but think the mine had been planted recently. Our presence was certainly not welcomed by all.'

Craig Egan, 9th Contingent

'You essentially had to treat the whole of Afghanistan as one big minefield in terms of moving around and being concerned for your own safety. The times that I got most concerned is when two vehicles would pass each other on a single lane road. One vehicle would have to move over to the side and lord knows what 'might be on the verge of a track.'

Harry Jarvie, Technical Advisor

'Harry Jarvie and I were visiting a survey team working near Gardez when we heard an explosion about a kilometre away. One of our guides relayed that a local boy had been severely injured and we decided to go and try to assist him. We were led down a pathway and could see land mines on the surface on either side of the path. When we got to the boy his face had been 'sandblasted' to a pulp and his right forearm hung by a few sinews from his elbow. Harry gave the boy a morphine injection and we bandaged him up as best we could before sending him off in a utility to get to the nearest doctor. We later learned that the boy had died not long after reaching the doctor. But more concerningly, we learned that the boy and his younger brother had been told by their father to go and collect some land mines so that the family's crops could be protected.'

Greg McDowell, Technical Adviser

CHAPTER SEVEN
REST AND RECREATION

Work would take up most of your time and invariably you would commit more than normal working hours to work related activities when you were in Pakistan. Rest and recreation in Pakistan generally involved doing some fitness activity, hanging around the house, going out to explore some part of Peshawar like the old bazaar or socialising at the US Club. Trips outside of Peshawar up the Khyber Pass or to Darra required a bit more planning and preparation for permits.

When you were working in Afghanistan it was mostly a routine of working, eating and sleeping. When the opportunity arose however it was always of professional interest to visit a wrecked vehicle or plane or even an old fortress that had been liberated by the mujahideen. These forays were always inherently risky as they were sometimes some distance from the demining teams and in Afghanistan anything could happen. Additionally, the wrecked vehicles had not necessarily been cleared of their explosive ordnance. Many of the wrecked vehicles still had live ammunition stored in them. In some instances, mujahideen commanders had created 'stockpiles' of captured equipment and ammunition; but the items had been dumped off the back of a vehicle so it was all strewn on the ground and invariably dangerous. Once they learned that you were a professional soldier the mujahideen would typically invite you to take a few shots to demonstrate your prowess at handling a weapon—and it would have been rude to turn them down.

'Halfway through our posting we had a few days off from running courses. We had been reading history about the 'Great Game'. We mounted an expedition up the Swat valley to Gilgit and Chitral in the Hindu Kush. At the Karakoram Pass into China at 5540 meters (eighteen thousand feet altitude—same height as Everest base camp), I got out of our four-wheel drive and set off for a glacier only three hundred meters away. I was determined to get snow on my expensive mountain climbing boots reluctantly provided by Army Office. I got about fifty meters before running out of breath. I did make it to the glacier and back but now have a better appreciation about hiking at altitude. In Chitral, we were wandering through the streets when locals started turning up on their horses. We followed them to a rough dirt oval and witnessed a game of Buzkashi. Polo originated from Buzkashi which has been popular in the Hindu Kush for millennia. Buzkashi is similar to polo except that anyone who turns up with a horse can join in, there are no apparent rules, and the body of a goat is used instead of a puck. Sometime later, I was staying with the Australian military attaché to Pakistan and told him about our travels. He was mortified. Apparently, foreigners are kidnapped for sport in that area. Indeed, after Buzkashi it is the most popular sport!

The American Club was the only place in Peshawar where one could meet with fellow foreigners and have a drink. The manager was quite a character. If your bar bill was over a certain amount for the month you were awarded an honorary 'lounge lizard shirt'. Monte excelled himself and was the first awarded but there was significant competition. The club attracted a motley group. There were people working for the various NGO including World Vision. The manager of the World Vision program went on to take a leadership role in the International Campaign to Ban Landmines. The initiative was awarded the Nobel Peace Prize in 1997 'for their work for the banning and clearing of anti-personnel mines'.

The American Club was also home base for a number of skinny Americans who would mysteriously disappear from time to time. One individual, on his return from a month over the border, won the 'who can eat the most' competition. After demolishing a large number of Peshwari nan's and demoralising his competition, he moved on to dessert of even more nan. During the war, the CIA had provided Stinger anti-aircraft missiles to the mujahidin to combat Soviet helicopters in the valleys. With the Soviets departing, the mission was now to try and buy them back before they got into the hands of anti-American forces.

Drink of choice at the American Club was gin and tonic. With tonic being a malarial prophylactic, we were just taking our health seriously. Unfortunately, the tonic was not so good, so it had to be diluted with copious amounts of gin. This led to some unruly behaviour. Most notorious was tuk tuk racing. At the end of a night at the American Club, tuk tuk drivers would compete for our attention. After selecting the most competitive tuk tuks, it was necessary to race home. Leaning out in the corners similar to side car racing proved to be the fastest technique. Tuk tuks are not very stable when up on two wheels: best to keep them flat.

The Green Crescent hospital was across the road from the American Club. It was always busy given that there were four million Afghan refugees living in Peshawar. It became very busy when a battle was raging in Afghanistan. The first sign of a major engagement was trucks and busses barrelling down the road to the hospital. After one especially violent episode, we volunteered to give blood. The blood donating process seemed to go on for quite a long time. Afterwards we were all very lightheaded. One or two drinks afterwards we were cheap drunks. I expect that they took more than the prescribed litre of blood. We had our own catheters from memory, so that was ok. I expect if we hadn't brought our own, they would have just used what they had. God willing 'Inshallah' we would have been ok.'

Graham Costello, 1st Contingent

Members of the 1st Contingent enjoying a game of cricket against a UK side at Islamabad in 1989.

'I got back to the house one afternoon to find Shorty Coleman and Bob Kudyba steaming the explosives out of an RPG-7 warhead just outside my bedroom window. A lovely little charcoal fire with the warhead suspended on a wire above it. They just needed the marshmallows and all would have been dandy.'

Carl Chirgwin, 1st Contingent

'One of my tasks was to do a weekly admin run to Islamabad. The drive there and back was a nightmare. At the Aussie Embassy I sent of the weekly sitrep and collected any other signals for us. I also collected any mail and if I was able to, picked up some 'supplies' from the Commissary at the Canadian High Commission.'

Carl Chirgwin, 1st Contingent

'Some of us thought we would get some civvy clothes tailored to go back home with. When Monty Avotins' shirts were presented the stripes on the body were vertical, but the stripes on his shirt-sleeves were horizontal. It made for an unorthodox look.'

Carl Chirgwin, 1st Contingent

'We didn't have a phone at our house so each Friday I caught a tuk tuk into town to make a call back to the family in Oz. I waited nearly two hours for the operator to put the call through. And then after about four minutes the call dropped out...'

Carl Chirgwin, 1st Contingent

'There were a lot of people who claimed to be spies at the American Club. They were living Graham Greene fantasies. I also met Idi Amin's former personal pilot and the writer Robert Kaplan. It's funny who you run into at your local club.'

Paul Petersen, 1st Contingent

'There was lots of Soviet weapons and equipment at the smugglers bazaar. I've still got a Soviet Army watch I bought at the market. It works as badly today as it did in 1989. And vodka. There was lots of vodka.'

Paul Petersen, 1st Contingent

'I heard that a French nurse was stoned to death in Peshawar because she went jogging wearing shorts. Monty Avotins and Shorty Coleman went jogging in shorts a few days later, and they were stoned as well. They threw the rocks back! In fact, I think our people got stoned and shot at a few times for inappropriate attire. They were very serious about fashion in Peshawar.'

Paul Petersen, 1st Contingent

'We were impatient to visit the famous Khyber Pass. So impatient that we didn't wait for approval to go. After a very late night at the American Club, Carl Chirgwin, a Canadian officer and I drove (lights out) through the tribal areas towards the pass. We didn't get very far. Carl shouted 'Fuck my alligator!' as we hit a boom gate at a police post at 3.30 am. That was the first and only time I have ever heard that phrase, but it seemed entirely appropriate for the occasion.'

Paul Petersen, 1st Contingent

'Carl Chirgwin and I had one short 'holiday' during our deployment. With Arnie Palmer and a Canadian officer, we drove up the Hunza Valley to Gilgit and on to the Khunjerab Pass on the Chinese border. The Hunza valley the Aga Khan's

A view along the Karakoram Highway in 1992.

country and is one of several places that claims to be Shangri la. It was beautiful. The trip back to Islamabad wasn't. Sixteen hours in a bus at night with no noticeable use of the brakes. One passenger was so sick he had to be tied to the doorway 'half in and half out' so he could vomit freely as the bus sped along. It made the Grand Trunk Road look safe.'

Paul Petersen, 1st Contingent

'The UN Mission's security detail issued frequent warnings to members of the UN and NGO community members. One of the main warnings was not to visit certain areas of the city and to never travel around alone. Women were required to cover their heads and shoulders when going out and men to refrain from the wearing of shorts, singlets or short-sleeved shirts.

Taking photos in Pakistan was not an advisable business most of the time, anyone caught taking photos of bridges; government or military buildings could be arrested as a spy. Taking photographs of women was likely to start a riot and even taking photographs of men was fraught with trouble as the indignant subject was likely to demand payment 'Baksheesh' from the person with a camera.

Avoiding public gatherings was another rule, Pakistan had a very large calendar of public holidays and various disgruntled groups would use these days to gather in public places and start a demonstration that almost inevitably deteriorated into a riot of some form. I recall standing in the market place once and we were passed by a bevy of black shirted policemen touting a Lahti each. The Lahti was a long length of stout wood, like a tree branch, but very strong and was like a horsewhip. 'Pardon me, Sir,' said one of the policemen as they approached a group of local men, and commenced flailing them with their Lahti's. Crikey! Let's move on from here.'

Bob Kudyba, 1st Contingent

'The Smuggler's Bazaar was situated out in what was known as the Tribal Area, an area where the rule of law and so forth was not recognised. It was more like a Wild West town, where disputes were settled with a gun. Everyone walked around with a gun-everyone that is except us. We never went to this area alone, always in pairs and always with the driver, who would generally do the translating with the stallholders.

It was customary to begin any negotiations with a stallholder with a glass of tea, into which about 5 teaspoons of sugar had been placed. These small glasses were served on saucers and the tea was extremely hot.

Inside a shop at Smuggler's Bazaar in 1992.

A story went the rounds about a young Canadian lady who had gone to this bazaar and was lured down one of the alleys with a promise of what she was looking for could be found further down those alleys. She was never seen or heard of again and rumour had it she'd been kidnapped and taken to Iran for a ransom.

Anything and everything could be bought at the 'Smuggler's Bazaar', from pen-guns to knives, pistols and explosives such as military grade plastic explosives and gelignite, to whisky. Plus, the detonators and safety fuze needed to set it off. As the weather was quite hot, I would eye off all this gelignite sitting in stalls in plastic bags with some consternation. If you mentioned to the stallholder about the instability of gelignite when it crystallises, they would just give an indifferent shrug of the shoulders.

Also hanging up in most of the stalls were these black discs, suspended on string, about the size of a dinner plate. They resembled dried out meat, with a very sweet pungent aroma wafting off them. These discs were pure hashish, the locals would put this hashish into their water-cooled 'shisha pipes and smoke it. While alcohol was banned for Muslims, I am unsure about hashish. I once encountered these two blokes puffing on a big shisha pipe full of 'hash' and cradling their AK-47 rifles, one was fitted with a swing down type bayonet and the fellow had extended it to prod at a poster size caricature of President Najibullah.

If the prospective purchaser of a gun wished, they could check their purchase out by paying the stall holder some money for some ammunition and test it by just pointing the weapon skyward and firing it off. Obviously, the saying 'What goes up- must come down' was unheard of in the bazaar as this was a very common practice. The newspapers contained many stories of people being killed by this firing into the air and it was a major part of any big wedding, where, at the appropriate moment, all the males present with a gun (which usually meant every male), would let rip with a full automatic burst of fire into the air. About 100+ people a year were killed by this practice, but it didn't seem to bother anyone.'

Bob Kudyba, 1st Contingent

'In another incident at the bazaar, I was examining some items of explosive ordnance with another contingent member and two men emerged from the next stall; one with a pistol in his hand. These two men were talking excitedly amongst themselves and the bloke with the pistol flung his arm up in an arc and pointed the pistol skywards right next to us. He proceeded to squeeze off about 10 shots into the air. When the pistol emptied, they turned and went back inside. When the bloke flung up his arm, the pistol was pointed straight at both off us, if it had discharged then, we would have both copped a round or two.'

Bob Kudyba, 1st Contingent

'It was about this time that someone, or disaffected group, began planting bombs around Peshawar. Usually set off in cars in crowded places. These bombs were not very large, but did cause casualties. Some members of the contingent just missed being involved in one bombing, when they passed by a place just before a car bomb went off. The possibility of being in the wrong place at the wrong time was always present. We became quite concerned about these car bombs and any vehicles attempting to enter our house in Peshawar was subject to search, the boot, under the car and inside. It was about this time that the song by the British group 'The Dream Academy' came into my head: 'Life in a Northern Town' this was life in the town in the North West Frontier Province of Peshawar. The song still comes back to me now, nearly thirty years later. So much so, a bloke has it on his iPod.'

Bob Kudyba, 1st Contingent

'One time during a visit to the local Smuggler's Bazaar, we came upon a dispute between stallholders that had just occurred a minute or so before. One stallholder had become involved in a disagreement with another and one of them had produced an AK-47 and simply shot the other fellow to pieces with a full magazine of rounds. If we had come upon the scene two minutes before, we would have been caught up in the middle of it. The fellow who had been shot was thrown into the back of a pickup and carted off. There was blood everywhere and

Inside a machine shop at Darra in 1992.

the blood-stained pickup came back to the scene. It looked like someone had dropped 20 litres of red paint in the back of it and on the ground.'

Bob Kudyba, 1st Contingent

'Travel by vehicle was another hair-raising experience as there are little to no road rules in Pakistan, it is a case of everyone for themselves and the first man there wins. Leave a space and someone will drive into it, traffic jams and snarls are common. Our local drivers were not fazed by this and just pushed in there with horn blaring. Driving was done by two speeds-flat out and stop and at times you'd just close your eyes as the vehicle whizzed down narrow streets avoiding everything else on the roads by a mere whisker, the driver continuously leaning on the horn. I was travelling out to Risalpur one morning and the driver was going at a fair rate of knots when there was this tremendous 'bang' and a mangy looking dog cannoned off the front of the vehicle.'

Bob Kudyba, 1st Contingent

Snake charmer at Islamabad in 1992.

'All work and no play make Jack a dull boy as the saying goes, we did have a social life although this was confined in many respects as the local population didn't share our western tastes in music and so forth. Fortunately, all the contingents were permitted access to the US government employees club in Peshawar. The US Embassy had a club for its employees and the bloke running it was a real live wire when it came to organising functions. He was given to wearing those loud 'Hawaiian shirts' and was always active in the club organising something. One event we got roped into was a 'No Talent, Talent Night' that was a roaring success. It was also here that a lot of informal business was conducted with the other contingents and we managed to get things accomplished, such as getting explosive ordnance rendered inert. Relations with the other contingents were great; we had a good rapport with everyone, from the American and French contingents to the two-man Italian contingent.'

Bob Kudyba, 1st Contingent

'We got roped in to erect the newly constructed basketball goal posts at the American school. A piece of cake it was assumed as these had to be installed at a specific height. We arrived to install these one afternoon after a stint out at Risalpur and by then it was absolutely positively bloody hot out there. I recall these goals had been constructed out of 200mm water pipe and they were heavy, coupled with the heavy wooden backing boards and attempting to handle the steel pipe after it had been sitting in the sun all day was not so straightforward a task after all. The pupils at this school were all sitting there goggle eyed watching these five Aussies in slouch hats, footy shorts and nothing else attempting to stand these goal posts up and get them plumb and straight at either end of the court with constant yelps from touching the hot metal of the poles and cursing under the weight of the damn things eventually the task was completed and we retired to the shade trying to find out who it was who had got us 'volunteered' for the job.'

Bob Kudyba, 1st Contingent

'About this time, we arranged a farewell to the New Zealand contingent as they were returning home and a new contingent had come out to do a hand over with the contingent that had helped get us up and started. About this time also the Aussies received a challenge to a game of cricket from the British contingent, to be played in Islamabad on some days off. Both Australia and Britain maintained an Embassy in Islamabad and the British ambassador was proud to point out 'his bar' stocked those 750ml cans of Foster's Lager. The Australian Embassy made do with Queensland's favourite brew (at the time) of Fourex. Unfortunately, being an eight-man contingent, we were three players short-what to do? Help came our way when three of the Canadian contingent volunteered to step up and play with us. The Canadians actually did quite well with one of them a real mean slugger with the bat. Only one problem was they kept forgetting to take the bat with them when they hit a run. Reflex action was to throw the bat down and sprint like mad to the other wicket. I never laughed so much in a long time... 'The bat, the bat! Take the bloody bat with you mate!"

Bob Kudyba, 1st Contingent

'The US contingent were all Special Forces fellows. One weekend they asked me to act as a diving supervising officer so that they could maintain the currency of their diving qualifications. I agreed and we did this at the pool at the Pearl Continental hotel. Not exactly the toughest location.

Carl Chirgwin, 1st Contingent

'We were able to take advantage of some of these numerous public holiday periods to do some sightseeing. In one set we joined with the Canadian contingent and journeyed up to the Khyber Pass. My memories of this visit were our arrival at the Afghan/Pakistan border where a huge pair of ornate iron gates separated Pakistan with Afghanistan's border. We gathered at this spot for a few photos and the Afghan guards on the other side hastily ran up and closed the gates over. It transpired the Kiwis had lined up across the demarcation line and on the three count all took a pace back so they were standing in Afghanistan.

'Welcome to Afghanistan-Keep Right' said the sign on the Afghan side, while on the opposite side the sign proclaimed: 'Welcome to Pakistan-Please Keep Left'. So much history in that place as the Khyber Pass is where the armies of Alexander the Great passed centuries before and then the British as they passed through. The magnificent views of the Hindu Kush are certainly a sight to remember.'

Bob Kudyba, 1st Contingent

'I recall one night after being at the US Club that a very tired officer was being driven around from house to house by the tuk tuk driver. Eventually, one of the Chowkidars said 'yep, that one is mine' (or words to that effect) and he was ushered into bed.'

Anonymous, 1st Contingent

'Another day trip we made was to the town of Darra, here the entire town's main industry is the manufacture of weapons, every type of pistol and rifle imaginable is replicated by hand

The narrowest point of the Khyber Pass in 1992.

in Darra. From pistols to Lee Enfield rifles, they are all made here and test firing is only a few notes for a handful of rounds. In one stall a heavy Russian 12.5mm machine gun was set up on an AA mount and the shopkeeper showed its workings by shooting a few rounds up towards an overhanging cliff face behind the town's main street. I don't know if anybody was up there but I am sure they would have beaten a hasty retreat when the heavy Russian 'Dooshka' rounds impacted on the rocks up there.'

Bob Kudyba, 1st Contingent

Visiting the Old City in Peshawar in 1992.

'Peshawar's 'Old City' the original parts that had been its early history, were also fascinating although some of the narrow alleyways where the sun almost disappeared tended to make me uneasy about lingering too long in one place and also the massive overcrowding of humanity into buildings that looked way past their expiry date were a poignant reminder of how good we had it 'back home'. The main bazaar here was absolutely boundless and it was possible to buy anything from furniture to carpets to car parts to more guns and even green parrots in ornate wooden cages for sale. Then there was the area selling things like pots, kettles, teapots, ornate samovars and jewellery.'

Bob Kudyba, 1st Contingent

'Our visits to the nation's capital, Islamabad, were also welcome, the city resembled Canberra with its wide roads and endless round-a-bouts with a seething, endless mass of cars, trucks, buses, donkeys and carts, everyone with horns blaring and ever-moving. A rumour went around that the idea for Islamabad was taken from Canberra. The main places we visited were the Australian and British embassies and I think we once managed a visit to the American Embassy.'

Bob Kudyba, 1st Contingent

'One of the first things we noticed about Pakistan was the ubiquity of firearms. In the Federally-controlled provinces they were carried by the police and paramilitary security agencies. Although police in Australia are usually armed, the Pakistani police often carried longarms. Even the security guards employed by businesses often carried a Kalashnikov. When eventually we were able to venture into the Tribal Agencies, we saw every male over about 15 years of age carrying a gun. This was a reflection not only of the security situation, but also of a gun culture that had existed in the region for more than a century. An unfortunate impact of more than a decade of war in neighbouring Afghanistan had been a flood of sophisticated modern weapons into Pakistan, especially the area around Peshawar. This had flowed into the existing gun culture to worsen crime and violence in the region.

As part of our in-briefing process, the 1st Contingent arranged a visit to Darra Adam Khel (Darra for short), a famous gun-making and trading town that was then just inside the FATA, on the road to Kohat. To do this, we needed a permit issued by the Provincial authorities in Peshawar. After a drive of a more than an hour, we arrived in the main street of a bustling town, where we were told we would be met by 'Raymond.' Very quickly, an affable Pathan with excellent English came up to us and introduced himself as 'Rehman'—close enough. Rehman, it turned out, had worked during the 1960s as a gardener at the former U.S. signals intelligence facility at Badaber Air Base, near Peshawar. There he had learned the English which, polished over the years, now earned him a living as a guide at Darra.

Darra was astonishing. We saw local craftsmen hand-making copies of everything from Lee-Enfield rifles to Kalashnikovs and rocket launchers. The most sophisticated workshops had a lathe and perhaps a milling machine: the rest had files, hand drills and hammers. The youngest workers were boys less than 10 years old. Arms were copied down to the original makers' marks and serial numbers. In addition to the local small arms, the war in Afghanistan had added plenty of inventory at the larger end of the scale—120mm rockets, 107mm multi-barrel rocket launchers parked in rows awaiting sale, hand grenades, mines, etc. etc. The variety of weaponry was endless. Gunshots rang out constantly—we discovered this was a combination of buyers' test firing their prospective purchases and craftsmen proof firing the pieces they were working on. Given the ubiquity of gunfire, it seemed silly not to join in, so a couple of the team paid a few rupees to try out a Kalashnikov variant and a (fortunately authentic) Soviet 12.7mm DShK machine gun, from the middle of the bazaar.

We soon learned that there was another gun bazaar a lot closer to home. A little to the north of Hayatabad, towards Jamrud, was a market that sold, well, everything. Among expatriates this usually went by the nickname of 'Smugglers' Bazaar' because of the range of genuine and non-genuine illicit goods on sale, everything from cheap Russian watches to Mont Blanc pens. If they didn't have it, they'd get it for you. About the same distance to the south was a similar bazaar in the township of Bara. In those days the Bara market straddled the border with the Khyber Agency (part of the FATA) and once on the FATA side, the nature of the shops in the market changed quickly to include ones that sold guns and hashish. One of the hashish purveyors would even proudly show off the onsite production facilities, including hand-pressing of the hashish resin.

Having a test fire at Darra in 1992.

Bazaar became a must-see for any visitors, including curious do-gooders and the next contingent. Part of my handover to my counterpart from the 3rd Contingent, Bill Sowry, was an enjoyable visit that included a spot of Kalashnikov shooting in one of the shop yards.'

Andrew Smith, 2nd Contingent

'During the month of Ramadan, Muslims are not permitted to eat or drink during daylight hours. So, it's up at about 4.30 am for prayers, then an early breakfast before first light and then... nothing! No food, no smoking, chewing gum or drinking water all day, regardless of your work, until about 6.30 pm. At first, it seems quite easy to undertake, but is quite demanding as the last few hours really stretch your urge to eat. Towards the end of Ramadan, I was interested to see all the young camels and the variety of fat-tailed sheep in the streets, although I wasn't quite sure what this was all about. At the time, I had little knowledge of the workings of Ramadan and the following holidays of Eid. It was all very interesting to observe until I saw the Eid mass killing of all these young animals in the streets. The brutality of the treatment and the method of slaughter were disheartening even though I have now seen this throughout much of the Muslim world. It's quite a sombre thing to experience and really unnecessary, but it continues every year.'

Graeme Membrey, Technical Advisor

'Midway along the Grand Trunk highway, when travelling from Peshawar to Islamabad, you used to pass across a rickety iron bridge that was constructed in 1888, so the sign said. I remember the date clearly as it was the same year that my grandmother was born. Another, perhaps more important sign at each end of the bridge read 'Danger—maximum 10 vehicles only'. On multiple occasions I travelled across the bridge slowly and we often had to stop because of the congestion when drivers failed to give way of when allow others to pass. On such occasions there would often be up to 50 vehicles, including overturned trucks, stuck fast on the bridge for several minutes at a time.

Graeme Membrey, Technical Advisor

'The only place to get a decent feed in Peshawar was the Thames Restaurant. It became a bit of a favourite haunt. Just no alcohol to go with your steak. There were all sorts of interesting characters eating there. We heard that it was also a favourite with the senior mujahideen leaders.'

Brian O'Connell, 5th Contingent

'We decided as a team that we would celebrate Christmas and set about obtaining and decorating a suitable tree. A turkey was bought, prepared and cooked under the supervision of Dave Edwards who amongst his many other strengths, was a dab hand in the kitchen. We also decided that we would give the house staff and drivers Christmas gifts and pooled our cash. I can't remember all of the items but our objective was to keep it practical. I recall that prayer mats and frozen chickens were amongst the gifts provided. So, on the day the Pakistan staff were invited into the lounge around the Christmas tree. A brief explanation of the Christmas tradition was provided and the gifts handed over. I think the staff were genuinely pleased with our offerings. There were a lot of stories about how various contingents from other nationalities treated their staff and I'm proud and pleased to say that we had a good rapport with our staff and interacted with them well. At some point on Christmas Day our Pakistani cook approached a couple of team members and indicated that he would like some 'strong drink' to help us celebrate. Generally, this was not something we would do, but I guess the Christmas spirit was flowing in more ways than one. Whiskey was the man's drink of choice and he consumed it at the rapid rate. The situation got a little out of hand. Long story short we dispatched our van with two team members to retrieve his sleeping body from a roadside ditch covered in mud and vomit. We cleaned him up and put him to bed until he sobered. I think 'Baba' who was a venerable old Pakistani infantry man and head chowkidar (guard) would probably have horse whipped the cook given half a chance. Certainly, if the police had found him, he would have been caned. He later repaid our hospitality by raiding our rooms, stealing what cash he could find and then disappeared. He was an exception and the other staff could not have been more friendly, helpful and trustworthy.'

'Even though the Psychologist back in pre-deployment training raised living and working so close together over an extended period as a potential issue, we all seemed to get on quite well as a team and had a few good parties and movie nights. Brown Dog took on the role of movie selector and picked a wide range of films from some of the local video stores. Most were really good but I do remember a few 'barry crockers' [shockers]. At least he was mixing them up. There were also a few 'latest releases' available in Peshawar that were still showing at the cinemas in Australia. It appeared that someone had smuggled a video camera into the cinema and filmed the film as when the video started, the view would be adjusted getting bigger and smaller as the operator set up his video recorder and they would hold this for the duration of

The commander of Michni Post at the top of the Khyber Pass in 1992.

the film. You could even see people walking in late and taking their seat. It was just like being there! The movies were in English but had subtitles which looked to us like they were coming from Malaysia. The quality wasn't the best but they were still pretty enjoyable.

We also received a few videos from Australia that our families sent to us. We were only allowed to send letters and small packages through the diplomatic mail to the High Commission so sometime the videos got knocked back and returned to sender. This meant relying on the Pakistani postal system where there was always the risk that the packages would be stolen. My wife sent me a John Wayne movie which made its way to us and was highly regarded by the team so there was an order placed for a few more westerns. Sheree had watched the movie and taken the ads out, prompting Brown Dog to exclaim, 'You've got a top misses there two icky!' but the request was to leave the ads in during future movies as the team wanted to see what was happening in Australia. Perhaps a little bit of homesickness.'

Michael Kavanagh, 5th Contingent

'Phoning home was also difficult as there weren't many international telephone lines into and out of Pakistan so you had to book a call and wait for an international line to become open. I think there was a six-hour time difference between Peshawar and Sydney with daylight savings so around 2.00am was a good time to try and call home based on the time difference and the availability of international telephone lines. After you booked a call, you would have to wait until the operator called you back so we wait on the lounge in the TV room as you only had a few rings to pick up or the operator moved onto the next booking and you missed out on your call. Many times, the call from the operator wouldn't eventuate as no lines were available and it was not unusual to come downstairs in the morning and see someone asleep on the lounge still waiting for the operator to call them back. Sometimes, the call from the operator would come back three or four days after the booking was made. I remember booking a call on a Thursday or Friday night and on Wednesday or so getting a call back saying my international call was now available. When I advised I didn't want the call anymore because of the time difference, the operator was incredulous and not too happy with this picky Australian so I copped a Pakistani spray for that.'

Michael Kavanagh, 5th Contingent

'The Australian House also had a reputation from previous teams as throwing a good party and being a handy place to pop into for a beer. As such, we felt an obligation to continue this tradition and I would like to think that we maintained the high standard that had been set. The US Special Forces Team in the neighbouring house and the other expats involved in the Deming Program were frequent visitors with many a demining problem solved over a few beers. The UN Store in Rawalpindi and the contacts at the British High Commission were being stretched to keep up with demand.'

Michael Kavanagh, 5th Contingent

'The American Club in Peshawar was also a popular venue with the 5th Contingent. The Americans had a consulate in Peshawar that back in the day had apparently been, according to the rumours, 'overflowing' with CIA operatives and others working to bring about the downfall of the Russians in Afghanistan. With 'mission accomplished' following the Russian withdrawal, there weren't many people left in the US consulate when we were in Peshawar but there a couple of people who looked like they just stepped out of a Jason Bourne movie to make life interesting. The US club was upstairs in this little building on the edge of their compound in University Town with what seemed a 'reasonable' level of security but, with the benefit of hindsight about what was going on in the region at the time, was woefully inadequate. Once inside, you were required to buy tokens that you took to the bar to exchange for a beer. This was apparently required to get around the 'selling alcohol' regulations but we didn't really care as long as we could get a beer.

The American Club was also a place to meet some of the people working on the other UNOCA programs. I believe there were around 200 different programs going on at the time with demining just one of them. One of the strangest that I heard of was a group from Sweden or Norway or somewhere that was making concrete slabs that were intended as roofs for the mud huts in Afghanistan to keep the weather out. I remember asking if the walls were strong enough to support these concrete slab roofs and was advised that they were. Unfortunately, I heard later that sometimes they weren't and some walls had collapsed resulting in casualties.'

Michael Kavanagh, 5th Contingent

'One day driving through Peshawar we came across an oval with a game of Bushkazi in progress—an Afghan favourite. This is where two teams on horseback fight for a stuffed animal carcass with the aim of throwing or carrying it through a goal at either end of the field. The legs of the animal are usually in place so plenty of scope to pick it up and grab it off another player. There don't appear to be too many rules and not surprisingly it looked like it got a bit rough. A damn fine display of horsemanship though.'

Anonymous, 4th Contingent

'Getting home from the American Club was also a bit of an adventure. Tuk tuks were everywhere in Peshawar and were a cheap mode of transport. Even for westerners, the price was

A friendly game of bushkazi near Peshawar in 1992.

reasonable. I may have this wrong but I believe there were 17 rupees to the Aussie dollar back in the early 1990s so one rupee was about 5 cents and I believe it cost 100 rupees or about $AS5 to get home from the American Club. There were usually a few tuk tuks out the front of the American Club waiting to take patrons home and tuk tuk races from University Town to Hayatabad with slightly merry Aussie deminers was a common Thursday night event. We would usually tip the winning driver another 100 rupees just for the bragging rights.

After we had been there a few weeks, some of the guys said that the drivers had let them drive their tuk tuks and all you had to do was pay them a few hundred rupees more for the privilege. I was keen to try this out but no drivers would let me have a drive, even when I offered them a tip. Christmas Eve was a particularly cold night when I came out of the American Club and I approached a tuk tuk with the driver asleep in the back, which was not unusual, wrapped in his blanket that they wore as a coat. He turned on the little lights in the passenger seat and jumped in the driver's seat and off we went. I remember thinking how tough their life was in this part of the world and how good we had it in Australia so when we got home, I think I gave him 500 rupees and wished him a Merry Christmas, which probably wasn't too smart in a strict Muslim country. On New Year's Eve, I came out of the American Club and the same driver was there. He was happy to see me and I thought, this is my chance to have a drive. The driver said it was OK in his limited English, but only after we got out of University Town. So, once we were on the quiet roads he pulled over and let me have a go with him sitting in the passenger seat in the back.

I'd grown up riding motor bikes in the country so I had a rough idea what to do but the tuk tuks have a strange set up for the gears which are on the left hand handle bar. You pull in the clutch and turn the handgrip forward to go up a gear so by the time you are in top gear, which I think was 4th, the clutch is pointing at the ground. Anyway, off I went towards the Aussie House. I remember I went over a speed bump too quickly which the driver wasn't too happy about as I got a little bit of air on one of the back wheels but I hadn't gone far when we got stopped by a what I called a 'half section strength fighting patrol' comprised of four Pakistani Police Officers armed with rifles who were patrolling the area. They got us out of the tuk tuk and in their limited English wanted to know why I was driving. The driver was getting most of the attention and they kept looking in the back for some reason. I think I showed them my UN card and, fortunately, they let us go after a few minutes but that was the end of my tuk tuk driving. When we got home, I gave the driver 500 rupees for the experience and asked him what happened. With the help of our security guards, whose English was a bit better than most Pakistanis, it turned out the Police thought we were transporting drugs. This didn't make sense to me as the drug flow was from Afghanistan to Pakistan, not from Pakistan towards Afghanistan which was the way we were going. Either way, it was an interesting New Year's Eve.'

Michael Kavanagh, 5th Contingent

'During the Gulf War, the High Commission in Islamabad was trying to locate all the Australians in Pakistan in case things turned nasty and an evacuation operation was required. Colonel Cloughly (the Australian Defence Advisor at our High Commission in Islamabad) called us one day advising that there was an Australian lady who was living in Peshawar that the High Commission couldn't get in contact with and asking if we could contact her. It turns out she was running a pizza shop and had been in Peshawar for about 20 years or so. The guys working at Risalpur stopped at her shop one afternoon on their way home but were advised by the staff that she generally worked at night. So, after a couple of beers at the American Club, a few of us got in some tuk tuks and went to the pizza shop to see if we could locate this lady. When we got there, the staff advised that she wasn't there that night so we thought, why not have a pizza? While we were waiting for our pizza, this Pakistani man struck up a conversation with us. I remember thinking that he was quite well dressed and his

English was pretty good, which was unusual in that part of the world. After a while, he told us he was from ISI, the Pakistani Inter-Services Intelligence Agency, and that he had been following us. I remember one of us pointing out that the key part of being in intelligence was keeping a secret. He advised that he had been following us for a few days and we were 'doing the right thing' so he just wanted to let us know we were being watched and to keep doing the right thing. We finished our pizza, got the tuk tuks to take us home and the next day let the High Commission know that we had tried to contact the Australian lady without success and that a guy stating that he was from ISI let us know he had been following us. A few days later we were advised that the Australian lady had contacted the High Commission advising she didn't require any support. I don't believe we heard anything further on the ISI contact.'

Michael Kavanagh, 5th Contingent

'Smugglers Bazar was also an interesting place to visit. It was located on the western side of Peshawar just before the entrance to the Khyber Pass and you could buy just about anything from a toaster to an AK-47. It was located in the Tribal Territories which didn't appear to have much Pakistani control as drugs were openly available in commercial quantities. Hashish, which is a potent form of cannabis, was rolled into flat sheets, wrapped in glad wrap and hung in rows from the tops of the little shops which looked like timber packing crates. The hashish was black and looked like sheets of licorice but the smell was quite strong and not very pleasant.

Mines were also for sale and I believe prices started from $US5 for an anti-personnel mine and from $US20 for an anti-vehicle mine but these may have just been the westerner prices given there wasn't any shortage of mines over the border. We were receiving reports of PMN mines functioning on pressure release, usually when the rock that had been placed on them was removed, so we would check out the stalls and see what was for sale. We had been warned by Team 4 not to detail what types of mine we were looking for as this would result in an Afghan risking their life by going into a minefield to try and retrieve the type that we had asked about. As such we would just ask what mines they had and they would pull back the sheet covering the display cabinet to show us their wares. Fortunately, a pressure release PMN [mine] was recovered intact by one of the mine clearance teams with a little 'pimple' on top of the pressure plate which changed it from pressure functioning to pressure release. We had already placed an increased emphasis on the risk of pressure release devices within the training courses but it was good to see how it was constructed.'

Michael Kavanagh, 5th Contingent

'Carpet buying was also a popular activity. There were plenty of shops in Peshawar selling some magnificent carpets but there were also travelling carpet salesmen who would bring their carpets, and their team of assistants, to our house. If we were in a shop, we would be seated in a chair and offered a cup of the sweet Pakistani tea for the sales process, which could take some time. The process was usually the same each time with us advising the 'boss man' what size and colour we were after and their assistant would hold it up for our consideration. If it was a contender, it would be carefully draped over a chair or laid on the floor for consideration. If we didn't like it, it would be thrown into the corner with apologies for even showing it to you. This would continue until there were about 8 or 10 carpets for consideration and a huge pile of discarded carpets in the corner.

It was then time for final selection and haggling over the price. It would usually take at least 20 minutes to get to this point and more times than not, a price couldn't be agreed and the deal wouldn't be done so we would leave the shop and try again another day or start again at another shop, depending on how keen you were for a carpet. I had been told by the guys about Brown Dog's 'Roll Test' when he was buying carpets and I wasn't sure if they were having a lend of me or not but I went carpet shopping with him one day and it turned out to be true. In the final stages of negotiations, he would lie on the carpet on his back and wriggle around and advise that he didn't like the carpet because it wasn't soft enough. It really messed with the Carpet Merchants but it was a good way to leave when a deal couldn't be done. We also leant when they weren't keen to haggle that they had achieved their daily sales target. If they weren't interested, you would ask if they had made their target. Generally, they would respond with a big smile and a 'yes'. We also learnt this was usually because of a sale to an American as we found out they weren't very good at haggling as they thought it was impolite and not just part of the game. So, just to confirm, we would ask 'American pay too much?' and it was generally, 'Oh yes. Very, very too much'.so you would either need to start again at another shop or, more likely, just leave it to another day. Despite the process, we all seemed to pick up a few carpets during our time in Peshawar, with Bear O'Connell picking up some really impressive carpets.'

Michael Kavanagh, 5th Contingent

Pakistani children working in a timber furniture factory at Peshawar in 1992.

'We also managed a few R&R trips to Islamabad. After the 4th Contingent departed, it was only a month to Christmas so we were pretty busy settling into our roles so we generally stayed in Peshawar. Then the Gulf War started in January and the UN advised that we weren't even supposed to leave our houses, which we ignored. I remember going to a UN meeting in Peshawar with Bear O'Connell in early February where they were checking to see if we all thought it was safe to leave our houses again and their looks of horror when Bear advised that we had been working and travelling to Risalpur the whole time so it was probably OK for everyone else to do the same. So, it was mid-February by the time we first got to relax in Islamabad. A couple of us went down together and we were advised there was a Valentine's Day Party at the Canadian Club. That was really lame so we went to the American Club as they had a massive bowling alley with beers and pizzas so that sounded a lot better.

It was difficult to get into the American Embassy Compound as the Pakistanis had burnt the previous embassy to the ground in 1979, killing two Americans and two Pakistani staff members, so security was tight. To get in, you had to enter a bullet proof vault where you were locked in while a security guard behind the glass checked you out and decided if they would let you in or not. Fortunately, we were admitted so we made our way to the bowling alley and hooked into some beers and pizzas. A couple of young US ladies took a shine to Brown Dog and he appeared to be working his Australian charm to good effect, which didn't go unnoticed by some off-duty Marines who weren't too happy about this Aussie moving into their territory. At about this point, we decided it was probably best to check out the British Club. As mentioned, it was fairly easy to get in, at least compared with the American Embassy, so we had a good night there having a go at the Poms and their lack of sporting success. After things settled down following the Gulf War, we managed a few trips to Islamabad to let off steam with most time spent at the British Club.'

Michael Kavanagh, 5th Contingent

'About 3 weeks prior to departure I was involved in a bridge build accident at Puckapunyal that resulted in two badly damaged pointer and index fingers. This made the thousand handshakes required for good mornings at Risalpur training facility challenging but none the less I managed. Anyway, one night at the American Club I was talking to another deminer named Dave and noticed a real pretty shiela talking to her friends across the bar. I asked Dave about her and he told me she was an Indian/American girl that was over here as an NGO saving the whales or something like that. He also said she was rather stuck up and plenty of blokes had tried to woo her but failed miserably. Well, I told him she hadn't met anyone like this dashing Aussie lad before and after a few more drinks to lubricate the vocals I slipped in beside her at the bar and started up a conversation. Things seemed to be going swimmingly, I was chatting away getting a few laughs and the occasional hair flick, all the while Dave was within ear shot and directly behind her giving me the wink and thumbs up for encouragement. That's when it all went pear shaped. I had kept my right hand with its ugly black finger nails hidden by my side so as not to weaken my chances with this lovely lady but as the night progressed, I became more relaxed I forgot. I reached for my drink and she immediately cast her eyes upon the offending digits so in a flash I quickly stuck them into my jeans pocket only to have one of the nails get caught and flick up and hit her in the lip and land on the bar between us. We stood for what felt like an eternity starring at the blackened bloody nail when her eyes slowly raised to meet mine with a look of shear disgust. She promptly turned and walked away to the chorus of Dave's hysterics.'

Dean Brown, 5th Contingent

'From time to time we had the opportunity to seek some respite in Islamabad at either the Canadian or British Embassy bar or some unsuspecting Embassy workers house. Unfortunately, apart from a couple of individuals the Australian Embassy staff did not want much to do with us. The trip involved either catching a PIA flight (Inshalla Airlines) from Peshawar to

Islamabad or driving a couple of hours down the Grand Trunk Highway. Either option was not pretty. On this occasion Don Quick and Charlie Chan had gone to 'Slammers' for the weekend. They had planned to fly back to Peshawar, but due to a lot of 'celebratory fire' the aircraft returned to Islamabad. This then forced them to catch a taxi from Islamabad to Peshawar. Several hours later the taxi arrives at the Hayatabad house with Don and Charlie who were so white they almost glowed in the dark. Driving down the Grand Trunk during the day is bad enough, but they had driven it at night with most of the journey being completed without any headlights. For those who know Charlie he has olive skin, but it was not that night. Needless to say, the lads had a few drinks as soon as they arrived.'

Ben White, 5th Contingent

'Possibly the most frightened I have ever been was when I was a passenger in the front seat of a car travelling between Peshawar and Islamabad along the Grand Trunk Road. Firstly, we were in a right-hand drive car travelling on the right-hand side of the road. As the passenger on the left-hand side I would have to tell the driver when it was safe to overtake. The only problem was that there was little scope for overtaking and the road users ranged from massively overladen trucks, to massively overladen buses, to cars, to motorbikes, to tuk tuks to horses and carts. All of these clearly travelling at different speeds and many of them manoeuvring erratically to either avoid hitting or overtaking each other. Of the few times undertaking this hair-raising trip did we come a cropper by having to swerve off the road so as to not clean up a horse and cart that appeared unexpectantly in front of us.'

Marcus Fielding, 8th Contingent

'After a couple of months our beards were really starting to thicken up. No one trimmed them as that is not the culture/ fashion in this part of the world. So, we all went full 'Ned Kelly'. The only thing that became a little concerning was finding food scraps in your beard.'

Marcus Fielding, 8th Contingent

Marcus Fielding at Jamrud Fort near Peshawar in 1992.

'Unlike most western countries fireworks are readily available in Pakistan—mostly coming over from China. The range of fireworks available was vast and the prices cheap. So, what do several bored Aussie soldiers do on the weekend? —have a firework war! As we were living in two houses separated by a vacant lot of no man's land the stage could not have been better. An opening salvo of rockets usually got proceedings underway. The first round invariably involved the warring teams on their respective roof tops striving for the perfect rocket hit. There were occasions when the quality of the rocket left something to be desired and a 'misfire' would score home goals. Curiously one firework was shaped like a tennis ball and once the fuze was lit could be used grenade like—these were my personal favourite. If things got a little out of hand then a raiding party would sally across no man's land to mount a 'trench raid' with these grenades that dispersed a wonderful burst of shredded cloth. This was all good clean fun but sadly the debris left behind took some time to be cleaned up. Once the local police came around to see what was going on and then stayed to watch the proceedings.'

Marcus Fielding, 8th Contingent

'Peshawar in the early 1990s was an interesting town; locals, millions of refugees, the UN, mujahideen, Pakistani military and police, drug smugglers, arms traders, diplomats and spies. Most of the Western types ended up having a drink at the 'US Club'—a ramshackle den of iniquity—especially when the Australian deminers showed up. Beer and/or darts comprised the entertainment menu but the conversations were the most interesting feature. And since the CIA fellows weren't permitted to cross the border, they were very keen to talk to those who were—which at that time was only us. But it is amazing how many beers one Aussie deminer can draw out of a CIA agent in the course of an evening. Most nights after last drinks we each grabbed a tuk tuk and encouraged our drivers to race against each other on the 20-minute drive back home. Even though it is physically challenging after a couple of ales quite a few are known to have 'surfed' their tuk tuk home— that is standing up as the machine navigated the potholes and corners.'

Marcus Fielding, 8th Contingent

Marcus Fielding at the Khyber Pass in 1992.

'We lived in houses just a few kilometres from the township that marked the eastern entrance to the Khyber Pass—a place called Jamrud. I woke every morning to the view of the Khyber Pass in the middle distance. At Jamrud, a bustling transit and trading centre, the entrance to the pass was ceremoniously marked by a large arch in the form of a brick fort over the road. Sadly, the size of vehicles has increased since the time the arch was constructed and the inside edges of the arch wore numerous battle scars from transiting overloaded 'jingly' trucks. The road is relatively flat and straight for the first couple of kilometres but before long it starts to climb and then the turns and switchbacks begin. We pass by the occasional village with mud-walled compounds and goat herds. When the surrounding hills start to get a bit steeper, we start to notice outposts (little forts) on the hilltops. As we go along, they become more numerous and it is clear that they are located to be able to provide support to each other.

Turning one corner the huge red brick fort comes into view. Shagai Fort was constructed in 1927 and has been a manned garrison ever since. The Khyber Rifles are the force that secures the Khyber Pass and the surrounding area and their Headquarters is in the Shagai Fort. As the mountains get steeper, we pass a section where there are dozens of brightly painted military regimental crests mounted on plinths or in the rock. They are the legacy of both British and Indian

The massive Shagai Fort in the Khyber Pass in 1992.

A proud member of the Khyber Rifles in dress uniform in 1992.

Army units that have had the pleasure of spending time on garrison duties in the Pass. At one point the road narrows to a single lane and then passes between two natural walls—this is the narrowest point of the Khyber Pass—about 10 metres. More switchbacks, more elevation, steeper surrounding hills, more little forts and then ultimately you arrive at Landi Kotal—a surprisingly large town that technically marks the Western end of the Khyber Pass. Land Kotal is key strategic hub located on a flattish spot in the high mountains of the Spin Ghar range. It is dominated by a sizeable railway yard which clearly served as the terminus for supplies to British forces warring against the Afghans. A rail line runs from Jamrud to Landi Kotal but to navigate the climb it travels along a series of rail switchbacks and tunnels that are removed from the road way. We are led to believe that the Khyber Pass train still runs periodically but the prospects of seeing this mythical beast are slim. A few kilometres to the west of Landi Kotal where the hills once again become very steep is the Torkham border crossing point into Afghanistan where the Khyber Rifles maintain a strong presence.'

Marcus Fielding, 8th Contingent

Steam train about to head up the Khyber Pass in 1992.
An engine at either end to move along the switchbacks.

Regimental crests at the entrance to the Khyber Pass in 1992.

'Cannabis plants grow wild throughout Pakistan; mostly alongside roads and often several metres tall. The sticky resins of the fresh flowering female cannabis plant are collected by pressing or rubbing the flowering plant between two hands and then forming the sticky resins into a ball of hashish. Hashish is readily available throughout Pakistan and the FATA where it is sold by the 'brick' (about a kilogram usually wrapped in glad wrap). In my travels around the FATA and in Afghanistan we would often come across storerooms filled with bricks of hashish. As a token of friendship when visiting a shop, the shop keeper often gave you a nub of hashish to say welcome. Many local people, perhaps unsurprisingly, became hashish addicts and could be seen vacantly wandering around while stoned. These 'shish men' were a traffic hazard and were often clipped by trucks. The Peshawar police would collect them up and place them in a very hot shipping container for a couple of days to dry them out before releasing them. Of course, the big money though was in opium and throughout Afghanistan vast fields would be planted with opium poppies.'

Marcus Fielding, 8th Contingent

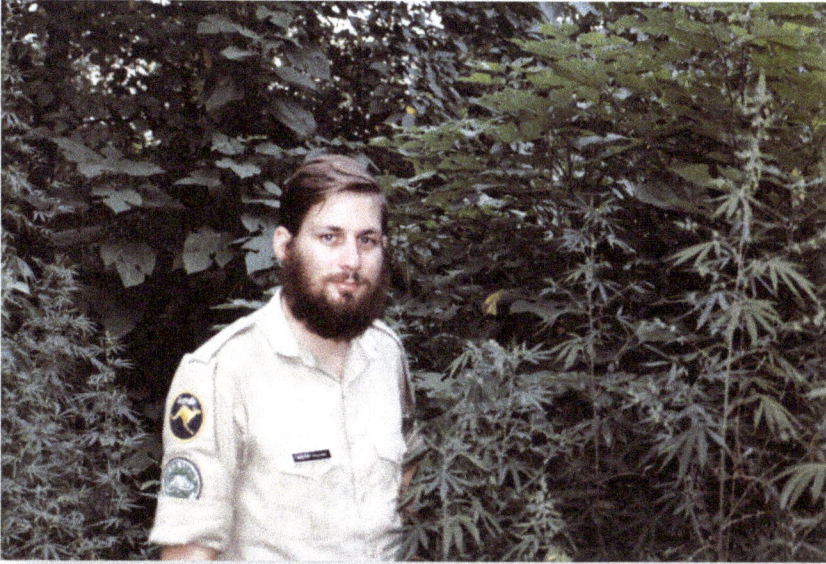

Marcus Fielding amongst a decent crop of cannabis plants
beside the road in Pakistan in 1992.

'It might sound a bit macabre and callous but lots of people were killed in that part of the world. Sometimes murdered but more commonly through accidents. Traffic accidents were big, as were accidental shootings, drug related deaths and electrocution. One day were read about a bus that had careered over an embankment as there was 35 dead and 60 injured; it must have been slightly overloaded...After a few months of reading about all of this in the morning newspaper we started to run a sweepstake over breakfast—who could guess the total number killed as reported on the front page.'

Anonymous, 8th Contingent

'After several months of work a few of us decided to take a few days off and go for a drive up to the Chinese border. The route would take us several hundred kilometres up through northern Pakistan and through some of the spectacular Karakoram Mountains. Our target destination was the Khunjerab Pass which is the highest border crossing in the world at an elevation of 4,880 metres. Completing the run down the Grand Trunk Road to a junction just before Islamabad in good order (i.e. no accidents) we turned

Darrell Crichton, Marcus Fielding, Clyde Jochheim and Danny Shaw at the Karakoram Pass border checkpoint between Pakistan and China in 1992.

north into the foothills and went through Abbottabad (where Osama bin Laden was later found to be hiding) and Manshera. Within a few hours we were on the Karakorum Highway and up above the snowline. Our 4WD vehicles were able to wind their way along the increasingly twisting and turning road through the valleys. We had our first long stop in Chilas and then stopped for the night at a 'truckers' roadhouse in Gilgit.

As we progressed the mountains became steeper and more majestic. The roads became windier and more 'hair raising' with precipitous drops down to the Indus River below and a surprising lack of barrier guards. Two cars passing by each other was scary enough but trucks and even small buses witnessed a big jump in the 'pucka' factor. The Karakoram Highway was carved through the mountains by the Pakistan and Chinese governments in the 1960s and 1970s. It was completed in 1979 and only opened to the public in 1986. About 810 Pakistanis and about 200 Chinese workers lost their lives, mostly in landslides and falls, while building the highway. A number of the highest mountains in the world are visible as you drive along the Highway.

Climbing ever higher the air gets thinner and the vehicles begin to labour. Suspension bridges allow us to cross the smaller valleys leading into the main valley; we are amazed that the roadway and the bridge are not connected to each other. One suspension bridge connects straight into a tunnel on the far wall of the valley. We pass through the villages of Hunza, Pasu and Khyber and wave to the locals. The way of life in these villages has probably not changed for centuries. Suddenly we seem to be so high that the villages stop and we seem to be on the home run to the Pass. The vehicles increasing gasp for oxygen and our ears pop. Eventually we drive into a relatively flat space and we have arrived at the Khunjerab Pass—the highest paved international border crossing in the world and the highest point on the Karakoram Highway. The Pass is often snow-covered during the winter season and as a consequence is generally closed for heavy vehicles from November 30 to May 1 and for all vehicles from 30 December to 1 April.

We stopped for a photo at the marker post indicating the border and to savour the moment engage in a brief snowball fight. But the thin air brings our exertions to a swift halt and we agree to just savour the view and the satisfaction of having made our

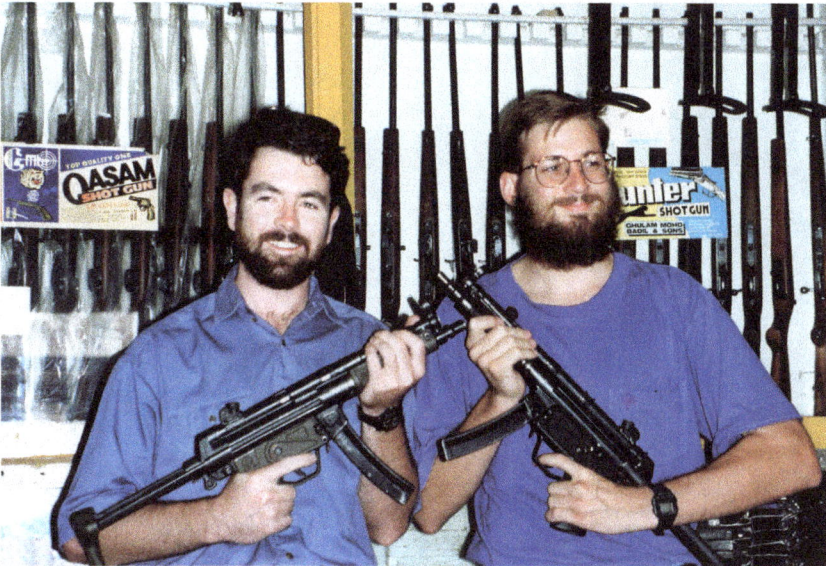

Mark O'Shannessy and Marcus Fielding at Darra in 1992.

target destination. Like summiting Everest, before long it is time to turn back and a couple of days later, we are back on duty in hot and dusty Peshawar.'

Marcus Fielding, 8th Contingent

'I recall going into the Pearl Continental Hotel. As you entered the foyer there is a nice sign reminding guests and visitors to check in their firearms with the concierge. I thought I would take the opportunity to get a haircut while I was there. The barber finished his task and gave me a lovely head massage before asking if I would like some oil in my hair. I thought—why not—something local and different. Big mistake—the oil wasn't 'light' and it took me about four shampoos to eventually get it all out.'

Anonymous, 8th Contingent

'Visiting a rug shop 'just for a look' was always an adventure. It didn't matter how much you were resolved to NOT buying something the proprietor would consider it a great failure of his skills if you didn't leave with something. Once the proprietor said take it as my gift knowing full well that your guilt will kick in and you will probably still end up paying him more than it was worth. The items we were most interested in though were the 'war' rugs which depicted a whole range of weapons and relayed several messages.'

Marcus Fielding, 8th Contingent

An Afghan 'war rug' depicting various weapons, mines and vehicles including helicopters.

'Landmine's are insidious things, made more sinister by the shape and size of some of them, which may be mistaken as a toy by children at play. Such was the case for Tariq, a 9-year-old boy who lost both hands. He was convalescing at a hospital in Quetta, in southern Pakistan. The ICRC staff who worked nearby invited me to visit, which I did as often as I could whenever I was in Quetta.'

Mark O'Shannessy, 8th Contingent

Mark O'Shannessy visiting Tariq, a mine victim, in Quetta in 1992.

'Towards the end of my tour I had to go to do a course in Kabul. We stayed at the UN house but it was in the period when the mujahideen factions were starting to get a bit testy about sharing power. So, a few rockets were lobbed at Kabul and we got a bit shaken about. The upside was that in the basement during one rocket attack I met my future wife who was also working for the UN.

Clyde Jochheim, 8th Contingent

Marcus Fielding, Mark O'Shannessy, Lee Uebergang and George O'Callaghan posing with two British motorcycle tourers at Peshawar in 1992.

'During a six-month deployment we were entitled to have a week of leave and a return flight to some destination paid for by the Australian Government. I can't recall where most of the others went—maybe Bangkok o back home? Danny Shaw and I spotted an interesting option and we had both left our run pretty late in the deployment anyway. The Iron Curtain was coming down and one of the local travel agencies was advertising a new Pakistan International Airways flight from Islamabad to Tashkent in Uzbekistan. We were both keen students of the Great Game so we thought—why not? Surprisingly the powers that be approved it all and so we spent a few days in Tashkent and then flew down to Bukhara for a few days. Tashkent was a typical Soviet city but experiencing central Asian people and their culture was very interesting. In Bukhara, an old Silk Road city, there was much less Soviet influence and many parts of the city looked like they hadn't changed for hundreds of years. We found a taxi driver who agreed to drive us anywhere over two days for US$100. He even took us back to his house and introduced us to his young daughters. I feel very lucky to have had the opportunity to visit this remarkable part of the world.'

Marcus Fielding, 8th Contingent

Guy Dugdale at the Khyber Pass in 1991.

'The first drive from Peshawar to the big smoke in Islamabad was a rite of passage early in any team member's tour. I do remember many newbies with eyes the size of saucers as they sat in the front seat of the Hilux van on their first trip to the big smoke! I was one myself, I think Mick Lavers may have been my guide.

One tale I recall on trip was a detour that was necessary as the highway was closed. The reason for the closure was a young man had self-immolated himself in protest that the local authority/council/government had withdrawn the bus service that would take students from the villages to a Polytechnic (trade school). His protest at this situation sparked a riot that lasted several days.

I often recall that story when people quip the often-posed question of why extremists (in particular Islamists) has such a low value of life. My experience was not that they placed a low value on life, but they were willing to offer themselves if they thought it would help their cause.

They had little education, no financial means and their lives were all they could offer. I suppose the bus service meant the difference to so many young men in the remote villages being able to learn a vocation and possibly break the poverty cycle, and to this young man his life was worth the sacrifice for others. I believe the bus service was restored after the riots ceased about four days later.

There are many other recollections of 'Jingly Trucks' rolled over due to poor loading and bad weather and locals reloading the cargo, often road base, by hand. I have fond memories of the samosas and coke served at the Indus Cafe located at the intersection of the mighty Indus and Ganges Rivers. It was about half way between Peshawar and Islamabad and became a regular and welcome constitutional break. Thanks Mick Lavers for the tip, not sure if others kept using the rest stop, but I certainly passed on the baton to others.'

Guy Dugdale, 9th Contingent

'We were in Jalalabad when a dog (Pronounced 'Spay' in Pashto) meandered through our camp site. It was healthier looking than most so I decided to call it over and try a game of fetch. I said, 'here boy, here boy...'. This went on for some time with no success, but generated much laughter from our Afghan driver Nazashar. As his laughter eased, Nazashar

Having a laugh with Afghan locals near Khost in 1992.

politely explained to me 'Mr Craig, you are a very funny man. This dog is Afghani, he does not speak English!' He had a point, so using my best Pashto I said 'Delta Raasha' to which, the dog promptly came over. Who would have thought? This was a simple but important lesson about unconscious bias, that has assisted me to transcend varied cultures, languages, and circumstances ever since.'

Craig Egan, 9th Contingent

'The occasions I enjoyed the most were our contingent gatherings at one of the Aussie houses. A simple barbecue, the good humoured practical jokes pulled by Billy Smith and Jim Horton, the entertaining stories by Mick Durnin, the sharing of an ale or two with all, and then to top the night off, a combined singalong (of sorts) to Tony Quirk's collection of 'Wallis and Matilda's' renditions of Banjo Patterson's poetry turned into song, such as 'Clancy of the Overflow.''

Craig Egan, 9th Contingent

'We had agreed to buy Afghan hamburgers for dinner one night while working in Jalalabad. Afghan hamburgers were delicious and consisted of deep-fried minced patties, a mixture of freshly chopped parsley and herbs, onion, and tomato all wrapped inside naan bread. Normally, we would let our Afghan teammates manage the purchase however over a long period our Contingent had become concerned about the trustworthiness of one our senior instructors. I decided to accompany him and observe the transaction. Upon being handed the change I asked our instructor how much it cost. To my surprise and alarm the shop owner took offence and proceeded to threaten me with a scolding of hot oil, which was resting in a ladle he held. Suffice to say, the senior instructor quickly escorted me from the shop and requested I remain outside with our driver. Naturally I obliged, understanding the cultural implications of Pashtunwali. We remained none the wiser as to the level of trustworthiness, and while I was bemused by the reaction, I thought a less obvious approach may be wise in future.'

Anonymous, 9th Contingent

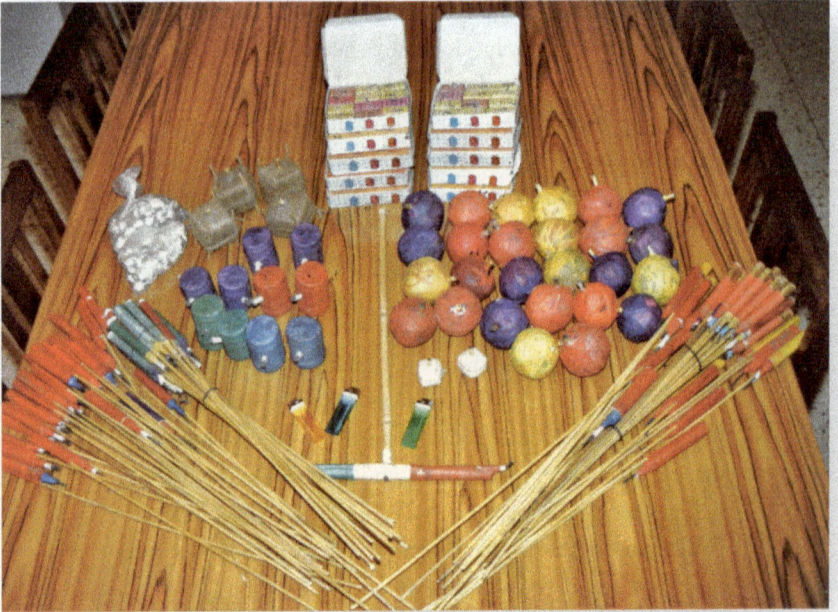

A collection of assorted fireworks.

'With fireworks readily available, and each Aussie house separated by a vacant block, the lure of fireworks battles was inevitable. Empty bottles would be aligned at varying intervals and angles along the exposed balcony of the second floor. Aimed at the neighbour's walled sanctuary, rockets would be launched without warning. Occasionally one would make the distance and the intended target, most would go astray, many would explode upon launch leaving the perpetrators ducking for cover. Two outcomes would follow, certain retaliation from the second Aussie House, and occasionally the local police got concerned that security had deteriorated in the neighbourhood.'

Anonymous, 9th Contingent

CHAPTER EIGHT
GOING HOME

The several days of handing over to the next contingent were typically very busy periods. The outgoing contingent had to try and cram as much knowledge as they could into the heads of the incoming contingent. Coupled with a touch of jet lag, a beer or two, and if you were unlucky your first dose of diarrhoea, the hand over period could be quite taxing on the incoming contingent members. And of course, the outgoing contingent members were all very much looking forward to going home. The adrenalin in both contingents gave the handover period a real buzz of energy.

As the contingent members all came from different parts of Australia the usual process was to fly back into Sydney and then catch a connecting flight back home. We travelled in civilian clothes of course. In those days there was no central debriefing sessions or even an official welcome home. Generally, only the senior officers were asked to provide a written report and perhaps a brief on that report to a more senior officer. Returning contingent members were required to do a medical and psychological check on return to their units but at the time these were pretty cursory.

'The team had done a great job building a professional demining course program. By the time we left the course materials were of the same standard as if the course had been run at the School of Military Engineering in Sydney. Local instructors had been trained up and thousands had been put through the courses run. The team had developed strong relationships with both the Afghan students and the Pakistan Army Engineers who hosted the training camp.

Being the first contingent, there was a chance that we might be able to nominate someone for an Australian honour or award. Every contingent member had done a great job. After a lot of consideration, it was decided to nominate one of the most junior contingent members. Consequently, then Bob Kudyba was awarded the OAM.

Our aim was to build a collection of ordnance and mines to be used in training back in Australia. They would be used to train future contingents. More importantly the collection would help prepare future deployments to other war zones (Cambodia, Timor and Afghanistan and Iraq since then). We had collected so much unexploded ordnance and mines that there was no way we were able to ship them home on a plane. Not only was the collection heavy, but despite our best efforts to remove all the explosives, they would have still triggered explosive detection alarms. A campaign was initiated to gain approval to acquire a shipping container to ship everything back to Australia. Eventually approval was given. Now what were we to do with all the empty space in a forty-foot shipping container? Contingent members took advantage of the empty space in the container to ship home some souvenirs. Arnie had a full formal dining table set custom made by the local craftsmen. The customs inspector couldn't believe his eyes when he opened the container in Port Botany. Fortunately, there was a few of us in uniform at the inspection to put his mind at rest.

Army Office had requested a briefing on our return. I flew down to Canberra. I was questioned by a room full of pasty-faced staff officers in a room without windows. I was surprised that they didn't seem to know anything about what the contingent had done. Despite our best efforts to send through copies of the training materials, videos and books created—they were lost somewhere in Army Office. Everyone was home safe, we had made the most of our opportunities, and a second contingent had been deployed—mission accomplished.'

Graham Costello, 1st Contingent

'On the journey back home, I got to see most of the team get off planes all the way down the east coast of Australia. As we had departed the country in a bit of a rush, there was no time for any Pre-Embarkation Leave, so the Army kindly gave it to us when we travelled back home. I was the last team member to get off a plane in Melbourne. I recall Peter Allen's song, as the plane circled over Melbourne and lined up on the runway to touch down at Tullamarine: 'I Still Call Australia Home', what a great feeling to come home and step out into the busy airport and its mass of humanity all going somewhere. There was no great fanfare or red-carpet welcome for me; I was just another face in the crowd. Nobody took much notice of me, nor seemed to care. The last vestiges of Operation Salam ended for me with the cheery voice of the very English flight attendant as I stepped off the plane: 'Do take care today!' she said with a beautiful smile. My first welcome was from my mother about two hours later: 'Where the bloody hell have you been?'

Bob Kudyba, 1st Contingent

The memento presented to contingent members as they completed their tours of duty.

'As we drew into late November 1991, I was informed that my replacement was being identified, though the specific officer had not yet been determined. Several Army colleagues, who were keen to take my place, had called and written to me asking about the conditions of service, the living conditions, the threats and so on. I was eager to go back home but somehow; I also wanted to stay and continue this line of work.

Graeme Membrey, Technical Advisor

'Before I left Peshawar, there were several functions to farewell me. These were all tremendously enjoyable affairs but the communal food plates proved to give me a parting gift. After a couple of weeks back in Perth I started feeling quite ill and visited the Army doctor. After doing a blood test he advised that I had contracted Hepatitis A! He explained that it came from food or water contaminated with faecal matter. He also told me that it affected primarily my liver and would give me yellow eyes and particularly yellow urine for many months. I was shocked and can only put it down to the food plates at the farewell functions.

Graeme Membrey, Technical Advisor

'The 6th Contingent arrived in late March 1991 and comprised 12 personnel. Bear O'Connell had advised Canberra that the team needed to be expanded following the departure of the US Contingent in early January for action in the First Gulf War. The 6th Contingent moved into the vacant US house which was diagonally behind our house. They were keen to get into it so we started handing over our roles. Captain Mick Lavers was the contingent 2IC and would take over my role as the Operations Officer on Demining Headquarters but Lieutenant Glenn Stockton would be the Administration Officer on Demining Headquarters as well as the Contingent 2IC. So I briefed up Mick and Glenn on what I thought the roles required, including a trip to Islamabad to introduce them to the Australian High Commission staff, get them registered at the UN store in Rawalpindi and, most importantly, introduce them to our friends at the British High Commission and their Australian beer as Ken had done for me.

We had a farewell party, which quite a few people attended, but I remember at one stage all the members of the 5th Contingent, including Ben White who was staying on as part of the 6th Contingent, made our way upstairs and we just sat with each other for a few minutes. I think it was at that point that it hit us that we were leaving and would soon be going our own ways. Perhaps that Psychologist who had briefed us during pre-deployment training was right. At the time, I thought he was preparing us for the issues that come from living so close together for so long, which he was, but perhaps he was also trying to prepare us for how close we would become as a team and how we would one day we would have to leave. We weren't together for long, Bear O'Connell said a few words and someone came up stairs and said 'what are you doing up here, the party's down stairs' and we went back down and got back into it.

Leaving our home in Peshawar was both happy and sad. Happy to be going back to family and friends in Australia but sad that our time in Peshawar was over. We flew PIA to Karachi and, as we had become acclimatised to flying in Pakistan, the poor state of the aircraft didn't surprise us. We then transferred to a Thai airlines overnight flight to Bangkok. For our return, we had a choice of taking a stopover in Bangkok, Hong Kong or Tokyo and we discussed it as a team and agreed on stopping in Hong Kong. I believe all the previous teams had stopped in Bangkok as it was the quickest way home but we were keen to see a bit more of the work while the Army was paying so decided on Hong Kong. We arrived in Bangkok early in the morning and transferred to our Thai airlines flight to Hong Kong arriving in the afternoon. The flight path took us over Vietnam and the Gulf of Tonkin and I remember looking down and thinking of all the Australians who had fought and died there as Vietnam wasn't the tourist hot spot it is now back in 1991.

When we arrived in Hong Kong, and they found out we had just come from Pakistan, the customs officials took quite a bit of interest in my luggage. There was nothing to find so we were off into Hong Kong for a bit of R&R. After checking into our hotel, it was off to McDonalds for a big feed of junk food. After hoovering it down, it was on to the first of many pubs. There were quite a few US Navy personnel in Hong Kong that night as they were returning home after the First Gulf War so we had plenty of drinking buddies.

The next morning, we were booked on a tour of Hong Kong city which had been included in our package, so I managed to get up and make my way to the lobby and found I was the only one from our contingent there. I purchased the obligatory tourist photo from the top of Mt Victoria looking a little shabby around the edges and was dropped off back at our hotel where the rest of the team had taken up prime position in the bar. Upon discussion of the night before, it turns out I was the first to call it quits, which is why I could make the tour, and some had kicked on well after I left. I think we had two nights and two days in Hong Kong before our overnight flight back to Sydney so I spent some time shopping and just decompressing from our time in Peshawar.

Before we boarded our overnight flight to Sydney, Bear O'Connell advised us that Major General Murray Blake would be meeting us at the airport to present us with our Australian Service Medals and that we had to be presentable so to take it easy on the alcohol. As a consequence, the flight back to Sydney was quite tame compared with some of our other flights.

After clearing customs and immigration, the five of us were ushered into a meeting room at the airport where our family members were waiting for us with Major General Blake, the Head of Corps and a few other military personnel. Major General Blake was an imposing figure, both physically and by reputation. He had served in Vietnam and was also about 6 foot 6 inches tall and looked like he could still run 5 km in under 18 minutes. He made a nice speech and congratulated us on our achievements before presenting us with our medals and I will never forget what he said, 'Wear this with pride. You have earned it and no one can take it away from you'. It resonated with me so much that I repeated these words when I had the honour of presenting medals later in my career.

Members of the 8th Contingent being awarded their Australian Service medals at Islamabad in 1992.

After the ceremony, we all said our goodbyes and went our separate ways. As three of us were from SME, we saw each other occasionally over the remainder of 1991 before posting orders sent us on the next stages of our careers. I believe that Browndog flew back up to Townsville that day. He was posted to the staff at Canungra where we caught up when I was there doing a course but, unfortunately, I never saw Barry Pickering again after we left that meeting room at Mascot airport.

Michael Kavanagh, 5th Contingent

'When we got home, we made sure that the single sheet of paper that had been stapled in our equipment issue record 'accidentally' got ripped off and lost. Sadly, that erased all evidence that we had ever been issued with some nice bits of kit for the deployment.'

Anonymous

'When I got home after six months away my wife was over eight months pregnant! I was very grateful to the other officers of the Regiment who had looked after her while I was away. Thankfully I was there for the birth of our first child.'

Marcus Fielding, 8th Contingent

'I returned home sporting a magnificent reddish-brown coloured beard that made me look like one of the singers from

ZZ Top. I was on leave so didn't have to take it off and thought it was a hoot to go into work and show it off to the others in the Regiment. Sadly, after a few days my wife 'convinced' me that perhaps it was time for the beard to go...'

Marcus Fielding, 8th Contingent

'I remember having to do a mandatory psych debrief a few weeks after we got home. I had an interview with a female psych. After a few minutes it became patently clear she had absolutely no idea of where I had been and what I had been doing—why would she I suppose. But after a while I got tired of trying to explain things to her so I just gave 'happy' answers to the questions she was asking and in 30 minutes it was all over. I left to get on with my life. Looking back now perhaps I should have taken the opportunity to have a better chat?'

Marcus Fielding, 8th Contingent

8th Contingent Members at Peshawar in 1992. Back Row from Left:
Barry Veltmeyer, Dean Beaumont, Darrell Crichton, Clyde Jochheim, Danny Shaw.
Front Row from left: Mark O'Shannessy, Marcus Fielding, Rex Wright,
George O'Callaghan.

The 8th Contingent welcoming the 9th Contingent at a BBQ at Australia House at Peshawar in 1992.

'We were deployed for over 5 months and didn't have much in terms of welfare support. A free 3-minute phone call home every week (or if disciplined enough, an accumulated 6-minute call every two weeks) = luxury! A letter in between those phone calls to supplement = luxury! And yet with this limited but appreciated welfare network, we still managed to plan a holiday in Hong Kong on the way home, for which my now wife flew to meet me, and we planned our engagement.

After returning home my fiancé and I went to the local supermarket. We got to the checkout when I noticed a large container of paprika and a massive bag of rice. These were removed rather swiftly, to the bemusement of the checkout staff, and much laughter from my fiancé. It took me six months to start eating either of these foods again, due to the perceived association with the 'recoilless' effect. Nothing wrong with me!'

Craig Egan, 9th Contingent

'Post Deployment Psych Debrief... 'No Doc, I'm fine thanks. I'm pleased with the way we managed some pretty intense and unpredictable circumstances and dangerous situations. Our training system is doing something right; we were well prepared and made calm decisions under immense pressure. I'm proud of our team, and our military. Our mission is making a difference. Being an unarmed mission helped lower our profile and blend-in, we were mot perceived as a threat, but it also created some significant vulnerabilities, especially when isolated and working in pairs. A few of my mates may like a bit of support—I will leave it to them—but I'm fine thank you.' End of Psych Debrief.

Now fast forward 10 years when my Officer Commanding and I are summoned to a brief with the Regiment's Commanding Officer. The Commanding Officer informs us that our Squadron may be deployed to Afghanistan to conduct demining operations. As such we were to prepare accordingly. It was during this brief I felt an intense rush of blood surge through my body, something I had never experienced before. My mind went into overdrive, 'You're not ... sending me back ... there!', 'That's' ... behind me now, I left it all there', 'I nearly got done-in last time, too many close calls!', 'Have we got any support?', 'Can we control the environment?'. All those memories buried deep in the back of my mind flooded forward. It took me another 10 years to seek help for my Post Traumatic Stress Disorder.'

Anonymous, 9th Contingent

The Taliban

The Taliban, alternatively spelled Taleban, which refers to itself as the Islamic Emirate of Afghanistan, is a Sunni Islamic fundamentalist political movement in Afghanistan currently waging war (an insurgency, or jihad) within that country. Since 2016, the Taliban's leader has been Mawlawi Hibatullah Akhundzada.

From 1996 to 2001, the Taliban held power over roughly three quarters of Afghanistan, and enforced there a strict interpretation of Sharia, or Islamic law. The Taliban emerged in 1994 as one of the prominent factions in the Afghan Civil War and largely consisted of students (Talib) from the Pashtun areas of eastern and southern

Afghanistan who had been educated in traditional Islamic schools, and fought during the Soviet–Afghan War. Under the leadership of Mohammed Omar, the movement spread throughout most of Afghanistan, sequestering power from the Mujahideen warlords.

The Islamic Emirate of Afghanistan was established in 1996 and the Afghan capital was transferred to Kandahar. It held control of most of the country until being overthrown after the American-led invasion of Afghanistan in December 2001 following the September 11 attacks. At its peak, formal diplomatic recognition of the Taliban's government was acknowledged by only three nations: Pakistan, Saudi Arabia, and the United Arab Emirates. The group later regrouped as an insurgency movement to fight the American-backed Karzai administration and the NATO-led International Security Assistance Force (ISAF) in the War in Afghanistan.

The Taliban have been condemned internationally for the harsh enforcement of their interpretation of Islamic Sharia law, which has resulted in the brutal treatment of many Afghans, especially women. During their rule from 1996 to 2001, the Taliban and their allies committed massacres against Afghan civilians, denied UN food supplies to 160,000 starving civilians and conducted a policy of scorched earth, burning vast areas of fertile land and destroying tens of thousands of homes. According to the United Nations, the Taliban and their allies were responsible for 76% of Afghan civilian casualties in 2010, 80% in 2011, and 80% in 2012.

The Taliban's ideology has been described as combining an 'innovative form' of sharia Islamic law based on Deobandi fundamentalism and the militant Islamism and Salafi jihadism of Osama bin Laden with Pashtun social and cultural norms known as Pashtunwali, as most Taliban are Pashtun tribesmen.

The Pakistani Inter-Services Intelligence and military are widely alleged by the international community and the Afghan government to have provided support to the Taliban during their founding and time in power, and of continuing to support the Taliban during the insurgency.

Pakistan states that it dropped all support for the group after the September 11 attacks. In 2001, reportedly 2,500 Arabs under command of Al-Qaeda leader Osama bin Laden fought for the Taliban.

Afghan Civil War 1996-2001

The Afghan Civil War 1996-2001 occurred between the Taliban's conquest of Kabul and their establishing of the Islamic Emirate of Afghanistan on 27 September 1996, and the US and UK invasion of Afghanistan on 7 October 2001.

The Islamic State of Afghanistan government remained the recognized government of Afghanistan of most of the international community, the Taliban's Islamic Emirate however received recognition from Saudi Arabia, Pakistan and the United Arab Emirates.

The defense minister of the Islamic State of Afghanistan, Ahmad Shah Massoud, created the United Front (Northern Alliance) in opposition to the Taliban. The United Front included all Afghan ethnicities: Tajiks, Uzbeks, Hazaras, Turkmens, some Pashtuns and others.

During the conflict, the Taliban received military support from Pakistan and financial support from Saudi Arabia. Pakistan militarily intervened in Afghanistan, deploying battalions and regiments of its Frontier Corps and Army against the United Front. Al Qaeda supported the Taliban with thousands of imported fighters from Pakistan, Arab countries, and Central Asia.

War in Afghanistan 2001-Present

The War in Afghanistan (or the U.S. War in Afghanistan or the Afghanistan War), code named Operation Enduring Freedom (2001–14) and Operation Freedom's Sentinel (2015–present), followed the United States invasion of Afghanistan[58] of 7 October 2001, when the US and allies successfully drove out the Taliban from power in order to dismantle al-Qaeda and to deny it a safe base of operations in Afghanistan.

Since the initial objectives were completed, a coalition of over 40 countries (including all NATO members) formed a security mission in the country. The war has since mostly involved US and allied Afghan government troops battling Taliban insurgents. The War in Afghanistan is the longest war in US history, having overtaken the US participation in the Vietnam War in 2010.

Following the September 11 attacks in 2001 on the US, which President George W. Bush blamed on Osama bin Laden who was living or hiding in Afghanistan and had already been wanted since 1998, President Bush demanded that the Taliban, who were de facto ruling the country, hand over bin Laden. The Taliban declined to extradite him unless they were provided clear evidence of his involvement in the attacks, which the US refused to provide and dismissed as a delaying tactic and then on 7 October 2001 launched Operation Enduring Freedom with the United Kingdom.

The two were later joined by other forces, including the Northern Alliance—the Afghan opposition which had been fighting the Taliban in the ongoing civil war since 1996. By December 2001, the Taliban and their al-Qaeda allies were mostly defeated in the country, and at the Bonn Conference new Afghan interim authorities (mostly from the Northern Alliance) elected Hamid Karzai to head the Afghan Interim Administration. The United Nations Security Council established the International Security Assistance Force (ISAF) to assist the new authority with securing Kabul, which after a 2002 loya jirga (grand assembly) became the Afghan Transitional Administration.

A nationwide rebuilding effort was also made following the end of the totalitarian Taliban regime. In the popular elections of 2004, Karzai was elected president of the country, now named the Islamic Republic of Afghanistan. NATO became involved in ISAF in August 2003, and later that year assumed leadership of it. At this stage, ISAF included troops from 43 countries with NATO members providing the majority of the force. One portion of US forces in Afghanistan operated under NATO command; the rest remained under direct US command.

Following defeat in the initial invasion, the Taliban was reorganized by its leader Mullah Omar, and launched an insurgency against the Afghan government and ISAF in

2003. Though outgunned and outnumbered, insurgents from the Taliban (and its ally Haqqani Network)—and to a lesser extent Hezb-e-Islami Gulbuddin and other groups—waged asymmetric warfare with guerrilla raids and ambushes in the countryside, suicide attacks against urban targets, and turncoat killings against coalition forces.

The Taliban exploited weaknesses in the Afghan government to reassert influence across rural areas of southern and eastern Afghanistan. From 2006 the Taliban made significant gains and showed an increased willingness to commit atrocities against civilians—ISAF responded by increasing troops for counter-insurgency operations to 'clear and hold' villages.

Violence sharply escalated from 2007 to 2009. Troop numbers began to surge in 2009 and continued to increase through 2011 when roughly 140,000 foreign troops operated under ISAF and U.S. command in Afghanistan. Of these 100,000 were from the US.

On 1 May 2011, United States Navy SEALs killed Osama bin Laden in Abbottabad, Pakistan. NATO leaders in 2012 commended an exit strategy for withdrawing their forces, and later the United States announced that its major combat operations would end in December 2014, leaving a residual force in the country.

In October 2014, British forces handed over the last bases in Helmand to the Afghan military, officially ending their combat operations in the war. On 28 December 2014, NATO formally ended ISAF combat operations in Afghanistan and officially transferred full security responsibility to the Afghan government. The

NATO-led Operation Resolute Support was formed the same day as a successor to ISAF.

At the beginning of Trump's presidency in early 2017, there were fewer than 9,000 American troops in Afghanistan. By early summer 2017, troop levels increased by about 50 percent; there were no formal plans to withdraw.

In August 2019, the US planned to negotiate with the Taliban to reduce troop levels back to where they had been when Trump took office. The Taliban remains by far the largest single group fighting against the Afghan government and foreign troops.

Tens of thousands of people have been killed in the war. Over 4,000 ISAF soldiers and civilian contractors, over 62,000 Afghan national security forces were killed, as well as over 31,000 civilians and even more Taliban.

CHAPTER NINE
REFLECTIONS

Memory is a curious and notoriously unreliable thing. Invariably all those who participated in the UNMCTT would have 'spun a good yarn' to others in the years after the mission, but none recorded their experiences. Some had maintained diaries but at the time there was a low level of interest in the experiences of those who had participated in the UNMCTT. Afghanistan was a little-known backwater and the Australian War Memorial's record stopped at 1975 at the end of the Vietnam War. The capture of anything to do with 'peacekeeping' activities by the ADF was not important so for many years our experiences become increasingly obscure. Even to the point that in the immediate aftermath of the 11 September 2001 al-Qaeda attacks on New York and Washington D.C. the ADF 'forgot' that they had previously had soldiers serving in Afghanistan less than ten years earlier. Nevertheless, now as we approach the thirtieth anniversary of the UNMCTT deployment I am pleased that these recollections and reflections can be captured for posterity.

> 'We arrived in Peshawar thinking we were the good guys and the
> Soviets were the bad guys. It wasn't like that at all. There were
> attacks and threats against Americans and other foreigners
> throughout our time there. It was clear that we weren't really
> welcome in 1989 and never really would be. I was surprised at
> the anti-American graffiti and other signs of hostility. It was

never clear what the Afghans and Pakistanis really thought of us. I saw it again in Iraq twenty years later, and I'm sure others have seen it since in Afghanistan. With hindsight, there was a lesson to be learnt by Australia in 1989, but we just didn't get it.'

Paul Petersen, 1st Contingent

'During its life span the UNMCTT was actually evolving into what the UN later created around the world so many times in so many countries—a Mine Action Program designed to address a country's coming out of conflict efforts to confront the issue of land mines and unexploded ordnance that will inevitably be a residual result of war. Little was I to realise (at the time) that I would see myself back working with land mines and explosive remnants of war in my first (of many) Mine Action Programs ten years later.'

Bob Kudyba, 1st Contingent

'In years to come Australians would also come back to Afghanistan as part of the United Nations Mine Action Programme for Afghanistan. Twenty years later, I would end up coming back to Kabul and Kandahar to keep working with the Afghan people to tackle the landmine and unexploded ordnance problem.'

Bob Kudyba, 1st Contingent

'When considering what I had learned from the Afghans, I realised how lucky I was to have been selected for that posting with ATC. If not for the cancellation of my other post in Iran and the efforts of my mentor, I might have spent the entire year back in Perth or Sydney. Instead, here I was after almost a year in this region of the planet which had brought so much to me.'

Graeme Membrey, Technical Advisor

'I brought back several habits from Afghanistan and my experiences there. The first is to treat your guests well. Give them the best of everything you have to offer. A second is the way of toasting drinks—by touching the lip of your glass below that of your compatriot as a mark of respect. A third is to remove my shoes before entering a house. And lastly, I learned to behave with humility. Curiously, our children all seem to have followed suit.'

Graeme Membrey, Technical Advisor

'I had a terrible run with gastro during my tour. I never could quite work out what caused it but for years afterwards I still wasn't shitting properly. There must be some really nasty bugs in that part of the world. In the end I had to get a full screen and took a course of drugs to try and burn it out.'

Anonymous, 5th Contingent

'I always look back with fond memories on the time I was fortunate to spend as part of the UNMCTT. I worked with some amazingly talented people such as Brian 'Bear' O'Connell, Dave Edwards, Benny Gardner, Glen Stockton, Dean Brown, Barry Pickering, Donny Quick and many more. I had the opportunity to return to Risalpur in the late 2000's and it is over-grown with those signature Australian gum-trees now taking a place of prominence. Nothing left and working with the Pakistani Army at the time I asked a very senior officer about the place and he spoke about it sadly with some disdain.

I don't think many of us realized at the time how dangerous the areas we worked in actually were and I think it is by good grace that we never had anyone seriously injured or killed over the entirety of the mission. It was a great adventure for many of us and as Australians tend to do; we waded through the bullshit and just got on with it.'

Ben White, 5th Contingent

'Incredibly, the UNMCTT mission wasn't recognised as being 'hazardous' in nature by the Australian Government until 2007—fourteen years after we had all returned! Thank fully it was backdated to June 1991.'

Marcus Fielding, 8th Contingent

'When al-Qaeda bombed the US embassies in Kenya and Tanzania in 1998 the US mounted a retaliatory strike into Afghanistan in response. Somewhere along the line my name had ended up in a file in the US Defence Intelligence Agency and as part of their planning I was contacted to provide advice on the land mine problem in Afghanistan should there be a downed pilot. I recommended that they contact the Demining Program directly for the most up to date information.'

Marcus Fielding, 8th Contingent

Marcus Fielding (centre) with Mohammed and Nasir Shah in Jalalabad in 1992.

'It was heartening to see that the Demining Program keep going after the UNMCTT guys left. Indeed, by the end of 1996 there were some 3,500 people in the program. Quite a few of the UNMCTT guys left the Australian Army and went back to work for the UN as civilian employees. Over time the tactics and techniques we developed in Pakistan and Afghanistan were exported to other mine affected countries like Cambodia and Mozambique. I understand that the Demining Program in Afghanistan continues to this day.'

Marcus Fielding, 8th Contingent

'Soon after the September 11 attacks it became clear that the US was going to go into Afghanistan. Not long after Australia announced that it would also contribute special forces troops. I thought surely all of us who have had experience in Afghanistan would be contacted to help advise those doing the planning or deploying. We weren't. This was particularly curious as the force commander was an Engineer and would surely remember that we had sent people there only eight years ago. After several weeks I suggested to a counterpart on exchange in the US that maybe this group could assist. He in turn raised a request on Australia from the US and we were then contacted and interviewed. It just goes to show that institutional knowledge doesn't last long.'

Marcus Fielding, 8th Contingent

'At the time, teaching Afghans how to employ explosives seemed like the right thing to do...until the US-led coalition invaded Afghanistan in 2001...'

Marcus Fielding, 8th Contingent

'The posting was a remarkable opportunity from a professional perspective as well as a personal and family experience. Being able to experience life in a Muslim and tribally oriented country traumatised by war and lacking in the very basic level of engineering services and law and order is sobering. To then experience hospitality and see the resilience of the individuals and families through years of war is humbling.'

David Taylor, Technical Advisor

'I was a student on one of the first demining course run by the Australian instructors. I remember them all and still stay in touch with Allan Mansell. I went on to become a demining instructor and later took on other more senior roles within the Demining Program. I have now set up my own demining company and do contract work. Over the years I worked with many Australians—particularly those who left the Army and joined the UN. I am very proud to have been able to make Afghanistan a safer place.

Haazrat Rahamany, Demining Instructor

'I was with a fellow UNMCTT veteran in 2011 conducting a handover in Afghanistan. We visited the weekly Afghan Bazaar/Market at Kandahar Airfield looking to purchase some local rugs. I commenced my first bartering transaction with a greeting using my very diminished Pashtu. In response, the Afghan gentleman asked where I had learned Pashtu? We explained our UNMCTT mission and the years of our separate deployments. He turned around and called out to several other stall operators' words to the effect 'Man from Mujahedeen time'. Our UNMCTT missions aligned with the great Mujahedeen struggle, therefore we appeared to be held in the highest regard by the locals.'

Craig Egan, 9th Contingent

Allan Mansell (right) with Haazrat Rahamany (centre) near Jalabad in 2005.

'Our Afghan instructors and drivers were critical to the success of the mission. We would provide overwatch and assuredness to the Program, they would deliver a majority of the training and most importantly navigate and negotiate our way across the complexity of language, culture, tribalism, corruption, and security in what was a warlord's domain. They enabled our missions, and I say with certainty 'saved' us numerous times, from capture, hostage, theft, and potential death. Without our Afghan teammates, our emergency funds may have been needed not only for additional supplies in extremis circumstances, but to leverage and influence a local warlord for protection and safe passage. Of note and for context, public whippings, hand removal, and hangings were a reality in Afghanistan at the time, and although the civil war had officially ended the warlords continued to challenge for supremacy and the spoils. This reality was the environment for which we immersed and thanks to our Afghan teammates, emerged. Respecting the language of their origins, to our Afghan teammates, I say Manana, Meraabani, Shukrya, Tashakor (Thank you).'

Craig Egan, 9th Contingent

After 1993, the mine clearance program in Afghanistan continued to mature and evolve. It remains the oldest and largest mine clearance programme in the world. The UN exported the hard-won knowledge gained from this mission to other locations around the world including Cambodia, Angola, Bosnia and Mozambique as part of its Peacekeeping Operations.

Since 1989 the land mine clearance program in Afghanistan has cleared more than 18 million items of explosive remnants of war (ERW), over seven hundred thousand anti-personnel mines, and more than thirty thousand anti-vehicle mines. A total of 31,860 hazardous areas have been cleared or otherwise cancelled. This represents 2,990 communities and over 2,850 square kilometres of land released.

Despite these achievements, the task of clearing all of Afghanistan is not yet complete and has been hampered by ongoing fighting. In 2018, some 1,415 Afghan civilians were recorded to have been killed or injured by land mines and ERW. Children comprise a significant percentage of that figure. Some 1,495 communities in Afghanistan remain threatened by mines, ERW and improvised explosive devices.

In 1997 the United Nations created the United Nations Mine Action Service (UNMAS) within its Department of Peacekeeping Operations. It was created to act as the UN focal point regarding mine action that specializes in coordinating and implementing activities to limit the threat posed by mines, ERW and improvised explosive devices. The term 'Mine Action' was chosen to encompass activities other than physical land mine and ERW clearance. The work of UNMAS is divided into five pillars of Mine Action—clearance, mine risk education, victim assistance, advocacy and stockpile destruction.

At the time of writing UNMAS provides direct support and assistance to 18 countries/territories/missions including Afghanistan, Central African Republic (MINUSCA), Colombia, Cyprus (UNFICYP), Democratic Republic of Congo (MONUSCO), Iraq, Lebanon (UNIFIL), Libya (UNSMIL), Mali (MINUSMA), Nigeria Palestine, Somalia (UNSOS) (UNSOM), Sudan, Abyei (UNISFA), Darfur (UNAMID), South Sudan (UNMISS), Syria and Western Sahara (MINURSO).

The ground-breaking work of the UNMCTT between 1989 and 1993 has gone on to become a world-wide movement to create a world free of the threat of land mines and unexploded ordnance.

United Nations Mine Action Service

The work of UNMAS is divided into five pillars of Mine Action—clearance, mine risk education, victim assistance, advocacy and stockpile destruction.

Clearance

In its broad sense, mine clearance includes surveys, mapping and minefield marking, as well as the actual clearance of mines from the ground. This range of activities is also sometimes referred to as demining.

Humanitarian mine clearance aims to clear land so that civilians can return to their homes and their everyday routines without the threat of landmines and unexploded remnants of war (ERW), which include unexploded ordnance and abandoned explosive ordnance. This means that all the mines and ERW affecting the places where ordinary people live must be cleared, and their safety in areas that have been cleared must be guaranteed. Mines are cleared and the areas are thoroughly verified so that they can say without a doubt that the land is now safe, and people can use it without worrying about the weapons. The aim of humanitarian demining is to restore peace and security at the community level.

Clearance methods:

Surveying—Surveying, or the formal gathering of mine-related information, is required before actual clearance can begin. Impact surveys assess the socio-economic impact of the mine contamination and help assign priorities for the clearance of particular areas. Impact surveys make use of all available sources of information, including minefield records (where they exist), data about mine victims, and interviews with former combatants and local people. Technical surveys then define the minefields and provide detailed maps for the clearance operations.

Maps—Maps resulting from the impact surveys and technical surveys are stored in an information management system, including a variety of programme databases, and provide baseline data for clearance organisations and operational planning.

Minefield marking—Minefield marking is carried out when a mined area is identified, but clearance operations cannot take place immediately. Minefield marking, which is intended to deter people from entering mined areas, has to be carried out in combination with mine awareness, so that the local population understands the meaning and importance of the signs.

Manual clearance—Manual clearance relies on trained deminers using metal detectors and long thin prodders to locate the mines, which are then destroyed by controlled explosion.

Mine detection dogs—Mine detection dogs, which detect the presence of explosives in the ground by smell. Dogs are used in combination with manual deminers.

Mechanical clearance—Mechanical clearance relies on flails, rollers, vegetation cutters and excavators, often attached to armoured bulldozers, to destroy the mines in the ground. These machines can only be used in certain terrains, and are expensive to operate. In most situations they are also not 100% reliable, and the work needs to be checked by other techniques.

Mine Risk Education (MRE)

Risk Education, or RE, refers to educational activities aimed at reducing the risk of injury from mines and unexploded ordnance by raising awareness and promoting behavioural change through public-information campaigns, education and training, and liaison with communities.

RE ensures that communities are aware of the risks from mines, unexploded ordnance and/or abandoned munitions and are encouraged to behave in ways that reduce the risk to people, property and the environment. Objectives are to reduce the risk to a level where people can live safely and to recreate an environment where economic and social development can occur free from the constraints imposed by landmine contamination.

RE, along with demining (which includes technical surveys, mapping, clearance of unexploded ordnance and mines, marking unsafe areas, and documenting areas that have been cleared), contributes to mine-risk reduction, or limiting the risk of physical injury from mines and unexploded ordnance that already contaminates the land. Advocacy and the destruction of landmine stockpiles focus on preventing future use of mines.

'Education and training' in MRE encompasses all educational and training activities that reduce the risk of injury from mines, unexploded ordnance and/ or abandoned munitions by raising awareness of the threat to individuals and communities and promoting behavioural change. Education and training is a two-way process, which involves the imparting and acquiring of knowledge, changing attitudes and practices through teaching and learning.

Education and training activities may be conducted in formal and non-formal environments: teacher-to -child education in schools, information shared at home from parents to children or from children to their parents, child-to-child education, peer-to-peer education in work and recreational environments, landmine safety training for humanitarian aid workers (Learn about the Landmine and ERW Safety Project) and the incorporation of landmine safety messages in occupational health and safety practices.

Victim Assistance

Building on the experience gained in this area since the entry into force of the Antipersonnel Mine Ban Treaty the negotiators of the Convention on Cluster Munitions agreed on a specific article on victim assistance (Article 5), which contains a number of obligations for States Parties with respect to cluster munition victims in areas under its jurisdiction and control. The Convention on Cluster Munitions also provides the following definition of cluster munition victims: '(...) all persons who have been killed or suffered physical or psychological injury, economic loss, social marginalisation or substantial impairment of the realisation of their rights caused by the use of cluster munitions. They include those persons directly impacted by cluster munitions as well as their affected families and communities.'

Hundreds of thousands of mine and explosive remnants of war survivors exist in 78 countries. According to the 2008 Landmine Monitor Report, there are up to 60,000 survivors in Afghanistan alone and over 45,000 in Cambodia. In 2011, the Landmine Monitor identified 4,286 new injuries around the world by mines, explosive remnants of war and victim-detonated improvised explosive devices. While the actual figure is unknown, it may well be far greater, since many incidents of mine and explosive ordnance accidents are never reported and are therefore not registered.

Within the UN system, the United Nation Mine Action Service works closely with the World Health Organization (WHO) and other UN entities, in particular UNICEF, that also support victim assistance activities. They all work closely with partner organisations outside the United Nations system, such as the International Committee of the Red Cross, Survivor Corps, World Rehabilitation Fund (WRF), Handicap International Belgium and Vietnam Veterans of America Foundation (VVAF).

Advocacy

UNMAS coordinates overall UN advocacy in support of treaties and other international legal instruments related to landmines and explosive remnants of war, including cluster munitions, and in support of the rights of people affected by these devices.

Public information dissemination—'Public information' in the context of mine action describes landmine and unexploded ordnance situations and informs and updates a broad range of stakeholders. Such information may focus on local risk-reduction messages, address broader national issues such as complying with legislation or raise public support for mine-action programmes. Public information 'dissemination', however, refers primarily to public-information activities that help reduce the risk of injury from mines and unexploded ordnance by raising awareness of the risk to individuals and communities, and by promoting behavioural change. It is primarily a one-way form of communication transmitted through mass media. Public information-dissemination initiatives may be stand-alone MRE projects that are implemented in advance of other mine-action activities.

Community liaison—Community liaison refers to the systems and processes used to exchange information between national authorities, mine-action organisations and communities on the presence of mines, unexploded ordnance and abandoned munitions. It enables communities to be informed about planned demining activities, the nature and duration of the tasks, and the exact locations of marked or cleared areas. Furthermore, it enables communities to inform local authorities and mine-action organizations about the location, extent and impact of contaminated areas.

This information can greatly assist the planning of related activities, such as technical surveys, marking and clearance operations, and survivor-assistance services. Community liaison ensures that mine-action projects address community needs and priorities. Community liaison should be carried out by all organizations conducting mine-action operations. Community liaison services may begin far in advance of demining activities and help the development of local capacities to assess the risks, manage information and develop risk-reduction strategies.

Stockpile Destruction

Stockpiled anti-personnel landmines far outnumber those actually laid in the ground. In accordance with Article 4 of the anti-personnel mine-ban treaty, State Parties must destroy their stockpiled mines within four years after their accession to the convention. Sixty-five countries have now destroyed their stockpiles of antipersonnel landmines, destroying a combined total of more than 37 million mines. Another 51 countries have officially declared not having a stockpile of antipersonnel mines and a further three countries are scheduled to destroy their stockpiles by the end of the year. There are many options available to states in destroying their stockpiles. Stockpiles are usually destroyed by the military, but an industrial solution can also be employed. The techniques used vary depending on the make-up of the mines and the conditions in which they are found. The complete destruction cycle involves aspects such as transportation and storage, processing operations, equipment maintenance, staff training and accounting, as well as the actual physical destruction.

Australian Members of the United Nations Mine Clearance Training Team

All ranks as at the time of deployment.

1st Contingent—July to November 1989

- 159884 Major Graham Ian Costello, Royal Australian Engineers (RAE)—15 Jul to 6 Nov 89.
- 321507 Captain Carl Gerard Chirgwin, RAE—15 Jul to 6 Nov 89.
- 434661 Captain Paul Gerard Petersen, RAE—15 Jul to 6 Nov 89.
- 1200254 Warrant Officer Class 1 Imants 'Monty' Avotins, RAE—15 Jul to 6 Nov 89.
- 255570 Warrant Officer Class 1 Alan 'Arnie' James Palmer, RAE—15 Jul to 6 Nov 89.
- 629783 Warrant Officer Class 2 Anthony Peter Smith, RAE—15 Jul to 6 Nov 89.
- 223917 Warrant Officer Class 2 George James 'Jock' Turner, RAE—15 Jul to 6 Nov 89.
- 227496 Staff Sergeant Craig 'Shorty' Andrew Coleman, RAE—15 Jul to 6 Nov 89.
- 320530 Staff Sergeant Robert 'Bob' Richard Kudyba, RAE—15 Jul to 6 Nov 89.

2nd Contingent—November 1989 to March 1990

- 313915 Major Willem 'Bill' Van Ree, RAE—27 Oct 89 to 11 Mar 90.

- 325386 Captain Bruce McEwen Murray, RAE—27 Oct 89 to 11 Mar 90.

- 282832 Captain Andrew 'Boomer' James Smith, RAE—27 Oct 89 to 11 Mar 90.

- 220875 Warrant Officer Class 1 Phillip James Palazzi, RAE—27 Oct 89 to 11 Mar 90.

- 1201492 Warrant Officer Class 1 Lesley Charles Shelley, RAE—27 Oct 89 to 11 Mar 90.

- 211890 Warrant Officer Class 2 Christopher Douglas Reeves, RAE—27 Oct 89 to 11 Mar 90.

- 221297 Warrant Officer Class 2 Graham Lionel Toll, RAE—27 Oct 89 to 11 Mar 90.

- 179190 Staff Sergeant Ian George Mahoney, RAE—27 Oct 89 to 11 Mar 90.

- 320311 Staff Sergeant Allan John Mansell, RAE—27 Oct 89 to 11 Mar 90.

3rd Contingent—March to July 1990

- 62665 Major Peter Anthony Kube, RAE—1 Mar to 29 Jul 90.

- 322498 Captain William 'Bill' Timothy Bolton Sowry, RAE—1 Mar to 29 Jul 90.

- 223426 Warrant Officer Class 1 Francis Joseph Duggan, RAE—1 Mar to 29 Jul 90.

- 553355 Warrant Officer Class 2 Noel Sean Channing Johanson, RAE—1 Mar to 29 Jul 90.

- 1205464 Warrant Officer Class 2 Ian Hugh Lawley, RAE—1 Mar to 29 Jul 90.

- 551523 Staff Sergeant Alan Kenneth Harwood, RAE—1 Mar to 29 Jul 90.

4th Contingent—July to November 1990

- 316118 Major Ronald James Morley, RAE—20 Jul to 3 Dec 90.

- 361264 Captain Kenneth Raymond Norman, RAE—20 Jul to 3 Dec 90.

- 222222 Warrant Officer Class 2 Danny John Hawkins, RAE—20 Jul to 3 Dec 90.

- 1205784 Warrant Officer Class 2 Bradley Mark Orreal, RAE—20 Jul to 3 Dec 90.

- 553506 Sergeant Kevin Desmond Darcy, RAE—20 Jul to 3 Dec 90.

- 48994 Sergeant Robert Paul Harbort, Royal Australian Infantry (RAInf)—20 Jul to 3 Dec 90.

5th Contingent—December 1990 to April 1991

- 1204059 Major Brian 'Bear' Philip O'Connell, RAE—22 Nov 90 to 5 Apr 91.

- 283249 Captain Michael Mervyn Kavanagh, RAE—22 Nov 90 to 5 Apr 91.

- 56862 Warrant Officer Class 1 David Christopher Edwards, RAE—22 Nov 90 to 5 Apr 91.

- 318559 Sergeant Michael John McQuinn, RAE—22 Nov 90 to 20 Jan 91.

- 553508 Sergeant Brenton Luke White, RAE—2 Feb to 9 Jul 91.

- 184080 Corporal Dean 'Brown Dog' Charles Brown, RAE—22 Nov to 5 Apr 91.

- 326280 Corporal Barry Thomas Pickering, RAInf—22 Nov 90 to 5 Apr 91.

6th Contingent—April to September 1991

- 57475 Major Brian Edward Gardner, RAE—27 Mar to 23 Jul 91.

- 230808 Captain Michael James Lavers, RAE—27 Mar to 7 Sep 91.

- 236723 Lieutenant Glenn David Stockton, RAE—26 Mar to 20 Sep 91.

Lieutenant Stockton was awarded a Chief of the General Staff's Commendation. The citation reads:

The Commanding Officer of the 7ᵗʰ Australian Contingent— United Nations Mine Clearance Training Team in Pakistan and Afghanistan has brought to my attention your exceptional performance as the Administration Officer of that Contingent.

Your dedication and professionalism in carrying out your administrative duties have contributed significantly to the success of the Australian Contingent's role in the United Nations Mine Clearance Program in Afghanistan. Furthermore, your perseverance and resourcefulness that you exhibited throughout your tour of duty resulted in you being held in the highest regard by members of the Australian Contingent, United Nations Officials and Diplomatic Staff with whom you worked.

I commend you for your performance of duty in this unique Australian commitment to United Nations international humanitarian endeavours. Your tireless efforts and pursuit of excellence brings great credit upon yourself, your Corps and the Australian Army.

- 224251 Warrant Officer Class 1 Christopher John Coles, RAE—26 Mar to 8 Sep 91.
- 316794 Warrant Officer Class 2 Michael Stuart Keen, RAE—26 Mar to 8 Sep 91.
- 2245510 Warrant Officer Class 2 Donald James Hayward, RAE—27 Mar to 23 Jul 91.
- 323806 Sergeant Adrian John La Fontaine, RAE—2 Apr to 8 Sep 91.
- 227909 Sergeant Joseph 'Joe' Lee Cochbain, RAE—27 Mar to 9 Jul 91.
- 226491 Sergeant Samuel Roy Snape, RAInf—26 Mar to 8 Sep 91.
- 553570 Corporal Charles Edward Gallagher, RAE—26 Mar to 8 Sep 91.
- 185120 Corporal Donald Alan Quick, RAE—26 Mar to 8 Sep 91.

7th Contingent—September 1991 to February 1992

- 211562 Major Warren James Young, RAE—1 Aug to 16 Dec 91.

- 228774 Captain Mark Neile Willetts, RAE—3 Sep 91 to 18 Mar 92.

- 182428 Lieutenant Guy James Dugdale, RAE—5 Nov 91 to 24 Mar 92.

- 122985 Warrant Officer Class 1 Raymond John Raddatz, RAE—3 Sep 91 to 16 Feb 92.

- 229413 Staff Sergeant Malcolm James Quigg, RAE—15 Jul to 14 Dec 91.

- 179147 Staff Sergeant Desmond Charles O'Hanlon, RAE—3 Sep 91 to 16 Feb 92.

- 226091 Staff Sergeant Wayne Bruce Schoer, RAE—3 Sep 91 to 16 Feb 92.

- 63719 Sergeant John 'Jungles' Patrick Kirkham, RAE—3 Sep 91 to 16 Feb 92.

- 229488 Sergeant Stephen James Charlesworth, RAInf—3 Sep 91 to 16 Feb 92.

8th Contingent—February to August 1992

- 226278 Major Rex Arthur Wright, RAE—6 Dec 91 to 6 Jun 92.

  Major Wright was awarded the Conspicuous Service Medal. The citation reads:

  For conspicuous service as the Officer Commanding the Australian Contingent United Nations Mine Clearance Training Team in Pakistan.

  Major Rex Arthur Wright is a member of the Royal Australian Engineer Corps who for the last six months has held the appointment of Officer Commanding the Australian Contingent United Nations Mine Clearance Training Team in Pakistan.

  On assuming command his energy, vision commitment and professionalism was immediately apparent to his subordinates, his staff and his superiors. He very quickly developed the impetus necessary to satisfy the high training standards essential for and inherent in the role of the Training Team. His dynamism, sense of duty and inspirational leadership, has ensured the achievement and maintenance of the highest possible state of training for the Afghan de-mining volunteers.

The Australian Contingent joined an international team of military experts in Pakistan in 1989 and is now the only remaining military contingent involved in the operation. While based in Pakistan to instruct Afghan demining volunteers for the majority of time, Major Wright's team have monitored and supervised missions to verify and enhance these activities is demonstrated by the increased number of Afghan refugees repatriating themselves to villages that were formerly uninhabitable because of the mine threat, and by the restoration of numbers of agricultural fields and grazing pastures which were previously minefields.

Major Wright's outstanding performance of duty in this humanitarian appointment has been of the highest order. His achievements have brought credit to himself, his Corps and Australia.

- 325341 Captain Marcus Conrad Fielding, RAE—11 Feb to 12 Aug 92.
- 233389 Lieutenant Mark 'MOS' Paul O'Shannessy, RAE—20 Mar to 5 Aug 92.
- 520509 Warrant Officer Class 2 George Patrick O'Callaghan, RAE—11 Feb to 12 Aug 92.
- 3143095 Staff Sergeant Clyde 'Jock' Byron Jochheim, RAE—11 Feb to 13 Aug 92.
- 74242 Sergeant Darrell William Crichton, RAE—11 Feb to 12 Aug 92.
- 3143101 Sergeant Barry Veltmeyer, RAE—11 Feb to 12 Aug 92.
- 629870 Sergeant Dean Lyle Beaumont, RAE—6 Dec 91 to 6 Jun 92.
- 1206118 Sergeant Danny Cedric Shaw, RAInf—11 Feb to 12 Aug 92.

During a visit to the mission in April 1992 Lieutenant General John Coates awarded the UNMCTT his Chief of the General Staff's Commendation. The citation reads:

I commend the Australian Contingent United Nations Mine Clearance Training Team for its exemplary performance of duty with the United Nations sponsored mission, Operation Salam. The Contingent's contribution to the training of Afghan refugees and villagers in mine awareness and clearance techniques has been of great humanitarian benefit to the people of Afghanistan.

The Australian Contingent joined an international team of military experts in Pakistan in 1989 and is now the only remaining military contingent involved in the operation. While based in Pakistan to instruct Afghan demining volunteers for the majority of this time, monitoring and supervising missions to verify and enhance mine clearance demonstrated by the increased number of Afghan refugees repatriating themselves to villages that were formerly uninhabitable because of the mine threat, and by the restoration of numbers of agricultural fields and grazing pastures which were previously minefields.

Australian Contingents have contributed much to Operation Salam's achievements and in doing so have brought great credit to themselves, the Australian Army and Australia. Their performance has been of the highest order and in keeping with the finest traditions of the Australian Army.

On completion of the Australian commitment to Operation Salam this commendation is to be returned to Australia and held in trust by Headquarters Engineer Centre on behalf of the Australian Army 'deminers'.

9th Contingent—August 1992 to January 1993

- 211713 Major Stuart Lee Uebergang, RAE—25 May 92 to 19 Jan 93.
- 1732454 Captain James Charlton Horton, RAE—3 Aug 92 to 19 Jan 93.
- 235873 Lieutenant Bruce Timothy Vivers, RAE—3 Aug 92 to 19 Jan 93.
- 222239 Warrant Officer Class 1 William Herbert Smith, RAE—3 Aug 92 to 19 Jan 93.
- 312382 Staff Sergeant David Robert Mitchell, RAE—3 Aug 92 to 19 Jan 93.

- 323902 Sergeant David Anthony Quirk, RAE—3 Aug 92 to 19 Jan 93.
- 182076 Sergeant Craig Douglas Egan, RAE—3 Aug 92 to 19 Jan 93.
- 514171 Sergeant Michael Jonathon Durnin, RAE—3 Aug 92 to 19 Jan 93.
- 226717 Sergeant Glyn Stephen Close, RAInf—3 Aug 92 to 19 Jan 93.

10th Contingent—January to June 1993

- 2222097 Major Ian Murray Bullpit, RAE—1 Dec 92 to 13 Jul 93.
- 5101618 Captain Craig Collis Thorp, RAE—1 Dec 92 to 13 Jul 93.
- 324669 Captain David Alexander Rye, RAE—11 Jan to 13 Jul 93.
- 2801302 Warrant Officer Class 2 Ross John Chamberlain, RAE—11 Jan to 13 Jul 93.
- 324357 Staff Sergeant Christopher Neil Reeves, RAE—12 Jan to 13 Jul 93.
- 230802 Sergeant Brian Clegg, RAE—12 Jan to 13 Jul 93.
- 551568 Staff Sergeant Phillip Murray Rowe, RAE—12 Jan to 13 Jul 93.
- 63949 Sergeant Craig David Crosby, RAE—12 Jan to 13 Jul 93.
- 325526 Sergeant Jay Craig Bruce, Royal Australian Army Medical Corps (RAAMC)—12 Jan to 13 Jul 93.
- 49694 Sergeant David John Black, RAInf—11 Jan to 13 Jul 93.

Technical Advisors

- UN Demining HQ in Islamabad
  - 314435 Lieutenant Colonel Ian Wallace Mansfield, RAE—1 Sep 91 to 15 Dec 92.

    When the UN Demining Program manager left the program in January 1992, Lieutenant Colonel Mansfield filled the position and was subsequently awarded the Conspicuous Service Cross. The citation reads:

For conspicuous service as the Commanding Officer of the Australian Contingent United Nations Mine Clearance Training Team in Pakistan.

Lieutenant Colonel Ian Wallace Mansfield is a member of the Royal Australian Engineer Corps who for the last twelve months has held the appointment of Commanding Officer of the Australian Contingent United Nations Mine Clearance Training Team in Pakistan.

On assuming command his energy, vision commitment and professionalism was immediately apparent to his subordinates, his staff and his superiors. This commitment has been infectious and greatly enhanced the United Nations humanitarian programs in Pakistan and Afghanistan. Furthermore, his dynamism, sense of duty and inspirational leadership, has ensured the achievement and maintenance of the highest possible state of training for the Afghan de-mining volunteers.

The Australian Contingent joined an international team of military experts in Pakistan in 1989 and is now the only remaining military Contingent involved in the operation. While based in Pakistan to instruct Afghan demining volunteers for the majority of time, Lieutenant Colonel Mansfield and his staff have monitored and supervised missions to verify and enhance mine clearance operations in Afghanistan. The success of these activities is demonstrated by the increased number of Afghan refugees repatriating themselves to villages that were formerly uninhabitable because of the mine threat, and by the restoration of numbers of agricultural fields and grazing pastures that were previously minefields.

Lieutenant Colonel Mansfield's outstanding performance in this appointment has complimented the earlier involvement of Australian Contingents and has brought great credit to himself, the Australian Defence Force and Australia.

° 223663 Lieutenant Colonel Gregory Colin McDowell, RAE— 15 Dec 92 to 1 Dec 93.

- Mine Clearance Planning Agency (MCPA) in Islamabad
  - ° 230808 Captain Michael Lavers, RAE—23 Sep to 14 Dec 91.
  - ° 378982 Captain John 'Jock' Francis O'Connor, RAE—4 Jan to 5 Aug 92.
  - ° 555358 Captain Harold James Jarvie, RAE—1 Jan to 5 Dec 93.

  Captain Jarvie was awarded a Land Commander Australia's Commendation. The citation reads:

  The Commander Australian Contingent has brought to my attention your exemplary actions in providing first aid to an Afghan boy injured by a mine on 23 August 1993.

  During a visit to a survey task in a minefield near Gardez in the Paktia Province of Afghanistan you witnessed a mine explosion. You accompanied a local along an unsurveyed lane through a minefield to render first aid to the injured boy.

  Upon reaching the injured boy, you reacted calmly and competently in spite of the boy's shocking blast injuries. You provided all possible first aid, administered morphine to ease the boy's pain, and facilitated his evacuation to the nearest medical facility. Tragically, the boy died from his injuries, but your actions did much to alleviate his suffering.

  I commend you for your courage in entering the unsurveyed lane through the minefield, and for your calm and competent handling of a critical first aid situation. In performing this act, you have brought great credit upon yourself, your Corps and the Australian Army.

- Afghan Technical Consultants (ATC) in Peshawar
  - ° 322518 Major Graeme Andrew Membrey, RAE—18 Jan to Dec 91.
  - ° 47195 Major David 'Paddy' Johnston, RAE—3 Jan to Dec 92.
  - ° 63340 Major David Taylor, RAE—Dec 92 to 5 Dec 93.

- South West Afghanistan Agency for Demining (SWAAD) in Quetta
  - ° 233389 Captain Mark Paul O'Shannessy, RAE—5 Aug 92 to Dec 92.

ACKNOWLEDGEMENTS

I would first and foremost like to acknowledge the other officers and non-commissioned officers who participated in this operation and who charted the pathway for the world to be demined.

Many of them continued to work in this field of endeavour with the Australian Army and outside of it after this operation. Some continue to do so some thirty years after their first exposure to this important work. Without their initiative and drive the demining program would not have been the global phenomena and success story that it is.

In particular, I would like to thank those who contributed 'snippets' as I called them—recollections or memories of their experience. In particular Graham Costello, Carl Chirgwin, Paul Petersen, Bob Kudyba, Andrew Smith, Allan Mansell, Michael Kavanagh, Craig Egan, Brian O'Connell, Michael Lavers, Mark O'Shannessy, Ben White, Bill Van Ree, Dean Brown, Warren Young, Glenn Stockton, Mark Willetts, Guy Dugdale, Ian Bullpitt, Dean Beaumont, Darrell Crichton, Clyde Jochheim, Lee Uebergang, Harry Jarvie, Greg McDowell, David Taylor and Haazrat Rahamany. Graeme Membrey generously allowed me to replicate snippets from his book *The Call of the Wild: Adventures and Near-Misses in 1991 Afghanistan*.

I would like to acknowledge the work of David Horner and the Official History team that produced the volume covering the UNMCTT and published in 2011. Their tireless and detailed research provided a tremendous record of the operation.

I would also like to thank my wife for her ongoing encouragement for me to write.

About the Author

Marcus Fielding was born and raised in Melbourne. He joined the Australian Regular Army in 1983 and graduated from the Royal Military College Duntroon as a Lieutenant in 1986.

In the following decades of military service Marcus held a broad range of senior appointments in Army, defence and inter-agency organisations in a number of locations throughout Australia and overseas.

Marcus has participated in four operational deployments. In 1992 he directed operations to clear land mines in Afghanistan. In 1995 he coordinated infrastructure construction projects in Haiti.

In 1999 and 2000 Marcus directed security operations and coordinated the repatriation of displaced persons as part of the Australian-led international force in East Timor. For his work in East Timor, he was awarded a Commendation for Distinguished Service.

In 2004 and 2005 he was the Commanding Officer of the 3rd Combat Engineer Regiment in Townsville—a unit of some 400 soldiers that formed part of the high readiness brigade of some 5,000 soldiers.

In 2008 and 2009 Marcus spent nine months in Baghdad as an 'action officer' in the Headquarters Multi-National Force–Iraq.

Colonel Fielding transferred from full-time to part-time service with the Australian Army in 2011. He now runs his own small business in Melbourne and is the President of Military History and Heritage Victoria.

INDEX